THE

GREATEST

HOAX

THE GREATEST HOAX

HOW THE GLOBAL WARMING CONSPIRACY
THREATENS YOUR FUTURE

U.S. SENATOR JAMES INHOFE

 WND Books

THE GREATEST HOAX

WND Books
Washington, D.C.

Copyright © James Inhofe 2012

Book Designed by Mark Karis

WND Books are distributed to the trade by:
Midpoint Trade Books
27 West 20th Street, Suite 1102
New York, NY 10011

WND Books are available at special discounts for bulk purchases. WND Books, Inc., also publishes books in electronic formats. For more information call (541) 474-1776 or visit www.wndbooks.com.

First Edition

ISBN 13 Digit: 978-1-936488-49-0

Library of Congress information available.

Printed in the United States of America.

ENDORSEMENTS

"The global warming religion is an aggressive attempt to use the climate for suppressing our freedom and diminishing our prosperity. Senator Inhofe's book is an important contribution in the fight against this totally erroneous doctrine which has no solid relation to the science of climatology but is about power in the hands of unelected global warming activists. Senator Inhofe is, in this respect, one of the world's most influential voices."

—The Honorable Václav Klaus, President of the Czech Republic

"Around the world, an army of many thousands of independent scientists disputes that human carbon dioxide emissions are causing dangerous global warming. Since making his first Senate statement about climate change policy in 2003, Senator James Inhofe has risen to become a much-respected senior General in that army. His fascinating new book reveals a politician of great integrity and courage, who is determined to fight for the interests of the general citizen against the tidal wave of false global warming science and propaganda that engulfs us. Read this inspirational book not only to become better informed about climate change, but also to understand better the public-spirited motivation of a rare senior politician who stands for genuine protection of the environment, and against the economically irresponsible proposals of the false doomsters of the professional environmental movement."

—Professor Robert (Bob) M. Carter, Marine Geophysical Laboratory, James Cook University

TABLE OF CONTENTS

ACKNOWLEDGMENTS

KAY HAS BEEN TOLERATING ME for fifty-two years of marriage, and it's always been a challenge. It seems as if I have always been involved in some conservative crusade that has created a lot of public hostility. She has suffered the consequences of me being accused of conservative extremism before conservatism was fashionable. Exposing the hoax has been no exception, but she has stuck with me and never complained. I love and appreciate my kids and grandkids whose friends had been brainwashed and have asked them why their father or grandfather was so extreme, but they never complained.

I appreciate Matt Dempsey, my Environment and Public Works (EPW) Committee communications director, who works tirelessly each day to expose the hoax. He and Katie Brown, my Environment and Public Works Committee press secretary, daily sift through a tremendous amount of material. They take it as a mission, and I so greatly appreciate their work.

I have had the benefit of many former staff who have helped me expose the hoax throughout the years. They have not simply helped me on climate issues, but they have helped me on a variety of conservative causes fighting overregulation and preserving free enterprise. I want to thank former EPW Committee staff director Andrew Wheeler who was with me for fourteen years. He and I fought alone especially in the early days of the climate change fight of 2003 to 2005. He did an extraordinary job helping me expose the hoax.

My former EPW Committee communication director Marc Morano, who had previously worked for Rush Limbaugh, regularly did an outstanding job exposing the hoax and gathering information from a variety of sources and researchers eager to be counted among the skeptics.

My EPW Committee advisors through the years included John Shanahan, Mike Catanzaro, and Tom Hassenboehler, who have helped me through the legislative meetings and debate on the Senate floor exposing the hoax and preventing the greatest tax increase in American history.

The foreign relations component to the hoax cannot be ignored. I appreciate my staff Luke Holland and Joel Starr for their traveling with me and their work.

My executive assistant Kathie Lopp has been with me for twenty years. She and my executive scheduler Wendi Price always keep me organized regardless of what conservative mission I have been on over the years.

My chief of staff, Ryan Jackson started with me when he was eighteen years old, and he's used to the battles we've faced over the years. Ruth Van Mark, my EPW Committee staff director, has been with me for twenty-two years. They work together to ensure my office keeps working efficiently.

I also appreciate the work of Jan Dargatz organizing a tremendous amount of material in multiple drafts over the past number of months.

My daughter Molly not only wrote my favorite chapter I call the "Igloo Chapter," but she helped with the entire manuscript.

Finally, I so much appreciate all my staff—past and present—who work hard with me each day for the people of Oklahoma and the nation. The staff deserves special recognition for the work they do.

WHY?

WHY?

Why, when the United Nations IPCC is totally refuted…

When Al Gore is totally discredited…

When man-made global warming is totally debunked…

When passing a global warming cap and trade is totally futile…

Why is this book necessary?

Very simple: the environmental activist extremists are not going away.

They are committed to their crown jewel—cap and trade. Whether by treaty, legislation or regulation, this largest tax increase in American history must become a reality for them.

And while the public has caught on and believes the global warming issue is dead, President Obama and the left in Congress are proceeding as if nothing has happened, inserting expensive CO_2 emission controls in virtually every piece of legislation and regulation.

While in the fall of 2011, the House, the Senate, and the Super Committee got all the public attention on a spending reduction of one trillion dollars over ten years, the extremists are pushing an agenda that would cost the American taxpayers three trillion dollars in that same ten years. And it has gone almost completely unnoticed for two reasons: one, the media is obsessed with resurrecting the Gore mantra and, two, MoveOn.org, George Soros, Michael Moore, and the Hollywood elites have the resources to reverse the defeat and preserve their crown jewel. And they will do it with or without President Obama.

This book constitutes the wake-up call for America—the first and only complete history of The Greatest Hoax, who is behind it, the true motives, how we can defeat it—and what will happen if we don't.

Somebody had to do it.

THE HOAX DEBUNKED: DON'T FEEL TOO SORRY FOR AL GORE

"With all of the hysteria, all of the fear, all of the phony science, could it be that man-made global warming is the greatest hoax ever perpetrated on the American people? It sure sounds like it."

—*Senator Jim Inhofe, July 29, 2003*[1]

SINCE JULY 2003, when I stood alone on the Senate floor and declared that man-made catastrophic global warming was the greatest hoax ever perpetrated on the American people, the credibility of the United Nations Intergovernmental Panel on Climate Change (IPCC)—which claimed to have a "consensus" on global warming—has eroded; cap and trade is dead and never to be resurrected, and, the belief that anthropogenic global warming is leading to catastrophe is all but forgotten. With the total collapse of the biggest campaign of his life, one might feel a bit sorry for Al Gore. Almost.

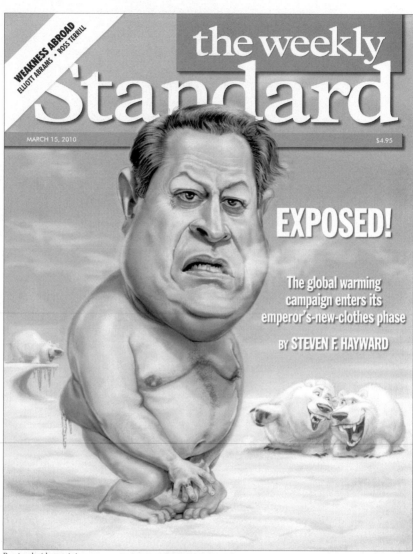

WEAKNESS ABROAD · ELLIOTT ABRAMS · ROSS TERRILL

the weekly Standard

MARCH 15, 2010

$4.95

EXPOSED!

The global warming campaign enters its emperor's-new-clothes phase

BY STEVEN F. HAYWARD

Reprinted with permission

When the Climategate scandal (which includes both the leaked Climategate emails and the errors in the IPCC science that were discovered in the wake of these emails) revealed that several of the world's top climate researchers were alleged to be cooking the science, Al Gore, the world's first potential climate billionaire was running for cover.

When he resurfaced months later with a nearly 2,000 word op-ed in the *New York Times*, the fact that he was in denial was clearly evident in the title of his piece: "We Can't Wish Away Climate Change."[2] I hated to break it to him, but global warming alarmism was long dead and buried at that stage. His op-ed was all about the so-called "overwhelming consensus," China, solar and wind power, globalization, rising sea levels, melting glaciers, and cap and trade—all topics that the American people no longer saw as relevant, especially in a weak economy.

What was Gore's take on the Climategate scandal? Climate scientists, he wrote in that same op-ed, were "besieged" by an "onslaught" of hostile information requests from climate "skeptics." He called those who question climate alarmism members of a "criminal generation," and said in a separate interview that these emails were just "sound and fury signifying nothing." Yet the *Daily Telegraph*,[3] one of largest publications in the United Kingdom, said it was the worst scientific scandal of our generation. The *Atlantic Monthly*,[4] the *Financial Times*,[5] the *New York Times*,[6] *Newsweek*,[7] *Time*[8] and many others, conceded that it was a legitimate scandal and that reform of the IPCC is absolutely essential.

The fact that *Time* magazine was one of the publications closely reviewing the Climategate scandal is important because this is the same publication that in 1974 told us that another ice age was coming, and we were all going to die.[9] And everyone remembers the *Time* magazine cover during the height of the alarmism campaign that pictured the last polar bear standing on the last cube of ice saying that we should all be worried, *very worried* about global warming.[10] So even a publication that has unabashedly promoted alarmism over both global cooling and warming was taking this scandal very seriously.

One of Gore's former global warming allies, President Barack Obama, has since received the message that global warming is no longer politically popular. Of course, the President was the one who sold his cap and trade agenda during the 2008 Presidential campaign as the only way to save the planet from catastrophe, promising that generations from now "we will be able to look back and tell our children that this was the moment... when the rise of the oceans began to slow and our planet began to heal."[11] He vowed that the United States would sign an international treaty with binding limits on greenhouse gases, and he promised that a cap and trade bill would be signed into law. But in a rare moment of clarity, then-

Senator Obama openly revealed the hoax for what it is, admitting, "if somebody wants to build a coal-fired plant they can. It's just that it will bankrupt them" and "under my plan of a cap and trade system, electricity rates would necessarily skyrocket."[12]

The central fact about cap and trade is inescapable: as the President said himself, it's designed to make the energy we use more expensive. Cap and trade would have been the largest tax increase in American history. It would have made electricity, food, and gasoline prices significantly more expensive, and it would have destroyed hundreds of thousands of jobs and further weakened the already fragile economy. The Wharton Econometric Forecasting Associates (or WEFA[13]) and Charles Rivers and Associates[14] found that the Kyoto Treaty would have cost the U.S. between $300 billion and $400 billion annually to implement. Cap and trade would have around the same pricetag. To put this in perspective, the Clinton-Gore tax increases of 1993, which were set at $30 billion, were the largest tax increases in American history. Cap and trade would have been ten times that amount.

What would Americans actually gain environmentally for all the economic pain they would be forced to endure? One would think that at the very least we would save the world from climate catastrophe. Not exactly. When Environmental Protection Agency (EPA) Administrator Lisa Jackson testified before the Senate Environment and Public Works Committee, she admitted to me that the United States acting alone on global warming would have no impact whatsoever on global carbon levels.[15] Administrator Jackson's comment echoes the work of Dr. Tom Wigley, one of Al Gore's own scientists, who said in 1997 that even if the Kyoto Protocol were fully implemented by all signatories, it would only reduce global temperatures by 0.06 degrees Celsius by 2050.[16] Such a small amount is hardly even measurable.

The results of such a plan are obvious: as jobs go to places like India, China, and Mexico, where they don't have any emissions requirements, much less the environmental standards we have in the United States, cap and trade would actually increase worldwide emissions. So in the end, it would have been all pain for no environmental gain.

President Obama no longer speaks about global warming, much to the disappointment of Al Gore, who in a June 2011 *Rolling Stone* article called "Climate of Denial" said,

> President Obama has thus far failed to use the bully pulpit to make the
> case for bold action on climate change. After successfully passing his
> green stimulus package, he did nothing to defend it when Congress
> decimated its funding. After the House passed cap and trade, he did
> little to make passage in the Senate a priority [...] The failure to pass
> legislation to limit global-warming pollution ensured that the much-
> anticipated Copenhagen summit on a global treaty in 2009 would
> also end in failure.[17]

But even though global warming hysteria and cap and trade are long
dead, the fight is far from over because President Obama is now moving
forward with a plan to achieve through regulation what could not be
achieved through legislation. In December of 2009, the Obama EPA
issued what is called the "endangerment finding"—a finding that greenhouse gases harm public health and welfare.[18] Armed with this "finding"
the EPA is planning to regulate greenhouse gases instead through the
Clean Air Act, which was never meant to regulate carbon. Like cap and
trade, this plan will have the same $300-$400 billion pricetag, it will put
the same amount of jobs in jeopardy, and it will cause the same amount
of havoc for our economy. My fight today is to stop them from achieving
this cap and trade agenda through the back door.

Unfortunately for Americans, the endangerment finding is just the
beginning. The Obama EPA now has the distinction of implementing
and overseeing the most aggressive regulatory regime in American history,
and, like cap and trade, it is aimed squarely at regulating traditional and
domestically abundant sources of energy like coal, oil, and natural gas out
of existence. The EPA is moving forward with an unprecedented number
of rules for coal-fired power plants and industrial boilers that have now
become known as the infamous "train wreck" for the incredible harm they
will do to our economy. They are set to destroy hundreds of thousands
of jobs, and significantly raise energy prices for families, businesses, and
farmers—basically anyone who drives a car, operates heavy machinery,
or flips a switch.

Indeed, from farm dust to puddles of water on the road, there are
very few aspects of American life that the Obama EPA is not planning to

regulate. And it is businesses and working families who will pay the price.

Meanwhile, President Obama also continues his administration's restrictions on deepwater permits in the eastern Gulf of Mexico and the Pacific and Atlantic coasts, and its constraints on development on federal lands. He repeatedly calls for increasing taxes on oil and natural gas producers, even though this will only serve to raise gasoline prices, destroy oil and gas jobs, and increase our dependency on foreign oil. His administration is also actively promoting the federal regulation of hydraulic fracturing, the primary method of extracting natural gas, even though states are efficiently and effectively regulating the practice, and there have been no confirmed cases of water contamination since the first use of hydraulic fracturing in my home state of Oklahoma in 1949. Well, we have seen the results of Washington's regulation on federal lands: it leads to less development, fewer jobs, and less economic growth. America has outpaced Russia to become the largest producer of natural gas because our immense shale deposits are located predominantly in areas of the country where states primarily regulate oil and gas development, not the federal government. In states like Pennsylvania, Arkansas, Oklahoma, Texas, Louisiana, West Virginia, Ohio, Michigan, and North Dakota, a boom in natural gas and oil development is transforming America's energy outlook—all thanks to the absence of federal red tape. Putting the federal government in charge of fracking will severely limit our ability to develop these vast resources.

President Obama has plenty of allies in this war on affordable energy who are working overtime to restrict our domestic energy supply. This "green team" includes the new Secretary of Commerce, John Bryson, who once called the Waxman-Markey cap and trade, a "moderate but acceptable bill"[19]; and his pick for Assistant Secretary for Fish Wildlife and Parks at the Department of the Interior, Rebecca Wodder. She is a staunch opponent of hydraulic fracturing, which she has said "has a nasty track record of creating a toxic chemical soup that pollutes groundwater and streams."[20] Also on the President's green team is his selection for chairman of the Council of Economic Advisers, Alan Krueger, who has made it clear that "the administration believes that it is no longer sufficient to address our nation's energy needs by finding more fossil fuels."[21] Perhaps the most telling comment in this war on affordable energy comes from Secretary of Energy Steven Chu, who told the *Wall Street Journal* in 2008, "Somehow we have to figure out how to boost the price of gasoline to the levels in Europe." That's interesting. Just what are those "levels" in Europe?[22]

- The United Kingdom: $7.87 per gallon

- Italy: $7.54 per gallon

- France: $7.50 per gallon

- Germany: $7.41 per gallon

So if you think paying over $4.00 is too much for a gallon of regular, fasten your seat belts. The Obama Administration is here to make it happen.

Meanwhile, Gore is still drowning in a sea of his own global warming illusions, desperately trying to keep alarmism alive. On September 15, 2011, Gore launched another global warming awareness event—a twenty-four-hour Climate Reality Project that was featured in several cities around the world and streamed live on the internet. But on that same day, in the midst of his latest campaign, Gore was dealt an inconvenient truth when the Obama Administration said that it would have to delay action on the EPA's greenhouse gas regulations, which were supposed to be announced September 30, 2011.[23]

The reason for this delay was obvious: not only would greenhouse gas regulations by the EPA cost hundreds of thousands of American jobs, they may cost President Obama his own job, and he knows that all too well. That's why he is punting on a number of EPA's rules until after the 2012 elections.

The reversal of the global warming campaign since the days when Al Gore and the IPCC both received a Nobel Peace Prize has been remarkable. In 2007, when the IPCC released its so-called "smoking gun" report, which claimed that the link between humans and catastrophic global warming was "unequivocal," it was met with a blaze of attention. On November 18, 2011, when the now discredited IPCC released the *Summary for Policymakers* for its latest global warming report, very few people even noticed. In fact, I was one of the only senators who weighed in on the report, as many of my colleagues were not even aware that it had been published.[24]

As for Al Gore, a September 2011 Rasmussen poll reported that "Despite winning a Nobel Prize and an Oscar for his work in the global warming area, most voters don't consider former Vice President Al Gore an expert on the subject."[25] Steven Hayward put it well in a March 15, 2010, article in *Weekly Standard* called, "In Denial: The meltdown of the climate campaign," when he said that Al Gore's *New York Times* op-ed was "the rhetorical equivalent of stamping his feet and saying, 'It is too so!' In a sign of how dramatic the reversal of fortune has been for the Climate Campaign, it is now James Inhofe, the leading climate skeptic in the Senate, who is eager to have Gore testify before Congress."[26]

In 2003, I stood alone in saying that the science behind anthropogenic catastrophic global warming was a hoax and that the so-called "solutions" were only symbolic and would have no impact on the climate. Now Gore stands alone in his dismissal of reform, openness, transparency, and peer-review to ensure good science, as well as his determination that man-made catastrophic global warming is a serious threat.

But you don't need to feel too sorry for Al Gore. Just as cap and trade was about to collapse, the *New York Times* reported on November 2, 2009, "Critics say Mr. Gore is poised to become the world's first 'carbon billionaire,' profiteering from government policies he supports that would direct billions of dollars to the business ventures he has invested in."[27]

Even without cap and trade, Gore had serious money-making potential from green government policies. As Stephen Spruiell put it in a March 22, 2010, article for the *National Review*, "Climate Profiteers: For Gore & Co., Green Is Gold":

> Only a small part of Gore's investment portfolio is tied to cap and trade. Most of the companies in which he invests would benefit from the other parts of the Democrats' energy bill—the parts that would be much easier for Congress to pass. Congress has been subsidizing green programs for decades, and that support increased dramatically with the 2005 energy bill. But the Democrats want to pump it up still more, even though the consensus for dramatic action on climate change is buckling like a shoddy roof in a blizzard of scientific scandals. The U.S. government, facing record-setting deficits and debt, cannot afford new subsidies. Yet with "green jobs" as their rallying cry, Gore and other advocates for more green-tech largesse will push to pick the taxpayers' pockets—lining their own all the while.[28]

Since then, Gore hasn't fared too badly. The *New York Times* article continues, "He has invested a significant portion of the tens of millions of dollars he has earned since leaving government in 2001 in a broad array of environmentally friendly energy and technology business ventures, like carbon trading markets, solar cells, and waterless urinals. He has also given away millions more to finance the nonprofit he founded,

©2010 by National Review, Inc. Reprinted by permission.

the Alliance for Climate Protection, and another group called the Climate Project, which trains people to present the slide show that was the basis of his documentary *An Inconvenient Truth*. Royalties from his book on climate change, *Our Choice*, printed on 100 percent recycled paper, will go to the Alliance, an aide said."[29]

Then there are his speaking fees—and he has been paid more than $100,000 for a single speech. "Mr. Gore's spokeswoman would not give a figure for his current net worth, but the scale of his wealth is evident in a single investment of $35 million in Capricorn Investment Group."[30] So he will be all right.

Don't feel too sorry for Al Gore. A billion dollars is a lot of comfort.

This book tells the story of my journey to halt the radical global warming agenda, which, at one time—when President Obama was elected and Gore was on top of the world—seemed inevitable. It began in 2003, when I was a one-man truth squad and culminated with my vindication when cap and trade was ultimately rejected, despite overwhelming Democrat majorities in both the House and the Senate, and when the Climategate scandal revealed allegations of cooking the science, which I had been saying all along. This was truly one of the most important fights of my career in public service because so much was at stake: as the leading Republican on the Environment and Public Works Committee and the father and grandfather of twenty kids and grandkids, I wasn't going to sit back and allow the largest tax increase on the American people to be imposed, all for nothing.

Despite what has been achieved, my story won't end at the conclusion of this book, much to the chagrin of my environmental friends, I'm sure. The Obama Administration will continue to pursue its war on affordable energy and, even if Obama is not reelected, global warming alarmists will continue fighting to implement their economically damaging agenda—and they have almost inexhaustible amounts of money and resources to advance it. MoveOn.org, George Soros, Michael Moore, and the Hollywood elite will still be going strong. But we will continue fighting back, and I believe the next chapter that continues after this book ends will offer many successes, too. In fact, we are already well on our way.

1

WHY I FIGHT

ONE OF MY GRANDKIDS ASKED me one day, "Pop I, why do you do things nobody else does?"

You see, "I" is for Inhofe, so it's "Pop I" and "Mom I," according to our twenty kids and grandkids.

My answer was simple, "Because nobody else does."

When I think back on those years when I was alone standing up against the global warming machine, I can't imagine what would have happened if I hadn't...certainly nobody else would. But I knew that I couldn't just stand by and let our government implement policies that would leave our country so much worse off for my twenty kids and grandkids.

Anyone who has visited me in my office knows how much my family means to me; they rarely leave without being bombarded with multiple pictures.

Even my readers are not exempt. For one thing, how many guys can actually say that they married the girl next door? I did, and we have been happily married for over fifty years, living in the same house in Tulsa that we moved into not long after our wedding day. Kay and I feel incredibly blessed by our four children, Jimmy, Perry, Molly, and Katy, and our twelve wonderful grandkids. My sons-in-law and daughters-in-law round out the twenty. Jimmy and his wife, Shannon, have three kids; Perry (a hand surgeon) and his wife, Nancy (a pediatrician) have two kids; Katy and Brad have three kids. They live virtually next door to Kay and me, and Molly and Jimmy with their four kids live only an hour away. How often does that happen? It usually doesn't, but it did.

Global warming activists often take the high moral ground and claim that they are on a crusade to save the planet for future generations. But what they never acknowledge is that their policies would give our children a substantially depressed quality of life, forcing them to live in a less free, less prosperous America. My dream for my children and grandchildren is the same as the dream of parents all over America: that our kids will reap the many blessings of living in a free country and that their opportunities will be even greater than our own. Under a global warming cap and trade regime, this dream would have just been a fantasy.

When I was ambushed by global warming advocates recently—no, they haven't given up—they asked me the same questions they always ask: "What if you're wrong?" and "If you're wrong will you apologize to future generations?" I always answer, "What if *you're* wrong? Will you apologize to my twenty kids and grandkids for the largest tax increase in American history?"[31] They usually don't have anything to say after that.

My willingness to take up the global warming fight when no one else would comes from so many experiences in my past as a father and grandfather, entrepreneur and public servant, that made me the guy I was to become when I finally took up the gavel as Chairman of the Senate Environment and Public Works Committee in 2003—a position that awarded me the best possible platform.

OVERREGULATION NATION

It all started in Tulsa, Oklahoma, with a fire escape—or more accurately, with the many ridiculous, unnecessary regulations surrounding that particular fire escape.

At the time, I was working long, hard hours as a developer. I was making fortunes, losing fortunes, expanding the tax base, doing the things I thought Americans were supposed to do. And all that time, the chief opposition I had to living out my American dream was, ironically, the government.

This was especially clear to me in the late 1970s when I purchased the Wrightsman Oil Estate, which had been abandoned for many years and had since fallen to ruin. In fact, it had become the favorite shelter for derelicts who were breaking in and burning the furniture to keep warm. But the house was extraordinary and it had a long, rich history. During World War II, the wealthy owners donated the mansion to the war effort, allowing the aviation community to live there while building the Douglas aircraft. Anyone who knows my love of flying would understand why I found its story so compelling. I saw a wonderful opportunity to restore the home to its former glory and couldn't wait to get started. I hired the construction crew immediately to expedite the renovation.

The only eyesore to the otherwise beautiful estate was one particularly ugly fire escape that was in plain sight. But I thought, no problem, I'll just move it to the north side of the building where it will still serve its purpose but not be visible from the street.

I was told that in order to do so, I would have to get the City Engineer's permission. So I promptly appeared at the City Engineer's office, made my request, and within moments I was flat-out rejected. "But I'm not changing anything to the structure or the foundation," I said, bewildered. "It won't create any safety concerns whatsoever. I'm simply moving the fire escape from the south side to the north side of the building. It still serves exactly the same purpose; it's just in a different location." But the Engineer was not moved. He told me I would have to take my request to the city board, put my name on the agenda, and they would hear my case in about two months.

"Two months?" I exclaimed, "This project can't be delayed that long because all the workers are being paid *now*. That will cost me thousands of dollars. Are you telling me you won't help me at all?"

He just looked at me and said, "That's your problem, not mine."

So I told him that I was going to run for Mayor and fire him. And I ran for Mayor and I fired him.

When I became Mayor of the City of Tulsa in 1978, I set out to make sure the size of the government, both in the operating budget and the numbers of employees, did not grow. I remember just after taking office a man came up to me and said, "Congratulations, Mayor on your victory. When would you like to have the Inhofe Hour, monthly, weekly, or daily?" When I inquired as to what the Inhofe Hour was, he said, "Well, we will let you go on our cable station to explain to the people your policies and programs." I responded, "You mean so that I can use the public funds to propagandize the electorate?"

The guy agreed that was a pretty good analysis, so I told him, "I don't want the Inhofe Hour monthly, weekly, or daily. As a matter of fact, I'm going to defund you." In the weeks after that, I found that they were running a script along the bottom of the screen of this cable station that no one watched, saying, "Your Mayor is trying to close the doors of government. Call the Mayor's office immediately and say that you demand to have this channel." We succeeded in defunding the agency and no one seemed to miss it.

Now that doesn't sound like a big deal, but it is. Most people just roll over when they are abused by big government. And bureaucrats seem to have unlimited resources—our tax dollars—to come after us. I hope you stick with this book to read a later chapter, the "Afterword." It will give you insight into the source of the power of the bureaucracy…(hint: bureaucratic earmarks.) And no one can appreciate the abusive government power until he has been through it. I've been there.

I never forgot how in my daily work as a developer I was bombarded with regulations from the city, state, and federal government, so much so that I once had to go to twenty-six different government bureaucracies in order to get a dock permit for a single condo project.

Having gone through that, I could understand why so many people just throw in the towel and close their businesses altogether. Unfortunately, this is exactly what happened to one of the best men I have ever known, my mentor A.W. Swift. When I was young I used to work with A.W. on his oil rigs, and he taught me so much about energy development and how to run a successful business. That was back in the days when you drilled for oil with cable tools as opposed to rotary bits. It worked on the principle of pounding. The bit would be raised up and dropped deepening the hole a few inches each time, until you struck oil—or didn't. It was backbreaking work, but I loved it and often worked two eight hour shifts without stopping.

A.W. had one son, Bert, who had become an engineer and the three of us worked together. One night, there was a terrible tragedy on the rig: the well that Bert and I were working on exploded and he was fatally burned. After Bert's death, I became like a son to A.W. I often thought of following in his footsteps and still wonder to this day what would have happened if I had made the oil business my career. But, of course, I took a different path.

About twenty years later, when I drove up to visit him at his home overlooking Keystone Lake, I was shocked to see all of his cable tool rigs stacked and sitting idle in his front yard. He told me sadly that he had given up the drilling business. When I asked why, he said, "It's because I can't handle the government regulations any longer. Here I am out working hard, trying to produce cheap oil for Oklahomans and the government keeps imposing regulations until they have just flat driven me out of business."

A.W. Swift was the epitome of the hard working, frontier spirited American. He was involved in an honorable profession, worked harder than most, was very religious, and provided jobs for as many as twenty-five or so people. Yet he was regulated out of business by the pseudo-intellectuals in Washington who think they know best. I remember thinking at the time how ironic it was that the very government that was supposed to create an environment where A.W. could achieve his American dream was the very institution that had managed to quell it. Something had to change.

"LONE VOICE IN THE WILDERNESS"
ONE-MAN TRUTH SQUAD: THE EARLY YEARS

Many think that my reputation for being a "one-man truth squad" on Capitol Hill started with the global warming debate, but actually, it started much earlier than that.

I remember back when I was first elected to Congress. The total national debt was $200 billion and I was outraged. I recall an ad on television that was produced by some national taxpayers group on what $200 billion was. They stacked up $1 bills in this 30 second spot until it reached the height of the Empire State Building, and that was $200 billion. Back then, I was a young, energized zealot who was going to do something about all of this. Come to think about it, I did do something about it.

When I was first in the Oklahoma Senate in 1968, then-Senator Carl Curtis, a Republican from Nebraska, asked me if I'd help him convince states to ratify a balanced budget amendment so that Congress would be forced to face up to the issue. Senator Curtis told me that when he annually introduced a budget balancing amendment to the Constitution, that the excuse used by his fellow senators for not voting for it, was that 3/4 of the states will never ratify it. So he charged me with the job of leading 3/4 of the states to "pre-ratify" the budget balancing amendment. I introduced and passed the first resolution by any state to join an initiative calling for a constitutional convention for the purpose of adopting a budget balancing amendment to the Constitution. We got to just one short of the 3/4 of the states necessary to pass resolutions calling for this constitutional convention. Then one of the national conservative groups decided that there was too much of a risk in calling for a convention, so the conservatives became split and the effort failed.

In a column called "A Voice in the Wilderness," Anthony Harrigan wrote, "Way out in Oklahoma there is a state senator who is going to balance the federal budget."[32] I've felt like that voice in the wilderness many times since then.

As a state legislator and as a mayor, there was only so much I could do to rein in the excessive regulations and out-of-control spending that were hurting our economy. The greatest overregulator of all was clearly the federal government, so I knew that if I wanted to be part of the solution, I had to go to Washington.

When I arrived in D.C. in 1987 as the Congressman from Oklahoma's First District, what surprised me the most was that so many of my liberal colleagues in the House, who were constantly pushing for more and more job-killing regulations, had never held a "real job" in their entire lives. Then I discovered that many members never go home. Why should they? There seem to be more golf courses in Northern Virginia than anywhere else. So I started the Tuesday—Thursday Club and went back and worked my district every weekend. Here's what happens: members who stay in Washington become part of the problem. You live next door to a lobbyist, a member, or a staffer. You see, there aren't any normal people in Washington and you forget what real people are like. Most of these guys don't understand the extent to which overreaching regulations are hurting businesses and job growth.

Much of that comes from the attitude of many on the liberal left that they know what's best for everyone else—an attitude that reached new heights during the global warming debate. D.C. politicians were working overtime to hide the real costs of their global warming cap and trade regulations because they knew the American people would never go for it. But as far as they were concerned, it was none of their business. They were going to save the world from global warming catastrophe whether their constituents liked it or not. They just didn't want voters to realize that this would mean the largest tax increase in American history, higher energy costs, hundreds of thousands of lost jobs, a depressed economy, and a much less free country.

I discovered that not only did my liberal colleagues in the House believe it was none of their constituents' business how they voted, but that there were set procedures in place to make sure of that.

There was one particular statement that I often heard whispered in the Cloakroom: "Vote liberal and press release conservative." Never once did I hear, "Vote conservative and press release liberal." I discovered that in 1932, a very powerful Democrat from Texas, Speaker John Nance Garner, had set up a system that would allow Democrats from conservative districts to do just that: vote for liberal causes, while pretending to be conservatives.

This is the way it worked: When a bill is introduced, it is assigned to a committee in the House of Representatives. In order for it to come to the Floor for a vote, there would have to be a committee hearing and a public vote. However, there was another more obscure method to bring it to the Floor for a vote, and that was for a majority of the Members of the House (218) to sign a Discharge Petition that was located in a drawer at the Speaker's desk. No member of the press or public could view the Discharge Petition, and no Member of Congress was allowed to have a list of the other Members who had signed a Discharge Petition. Of course, when you went to sign, you could see the other ten names on the page with you, but you were not allowed to write those names down. If any Member disclosed the names of anyone who had signed the Discharge Petition, the penalty was that the person could be expelled from House of Representatives.

This system seemed outrageous to me, and it was even more upsetting that no one had ever fought it. But it occurred to me that something could be done: I introduced a Resolution that would make the signatures of a Discharge Petition a matter of public record. Of course, the Speaker and the Democratic leadership didn't like that one bit, so they assigned my resolution to a committee that agreed never to bring it up for a vote. After that, the only way to get it out for a vote was to file a Discharge Petition on the Discharge Petition and have that placed in the Speaker's drawer. I knew that this would be very heavy lifting for just one Congressman, but I was determined.

Unfortunately, my Discharge Petition suffered the fate of the process I was trying to reform. So the only way I could get what I wanted was to expose the names of those Members who signed my Discharge Petition. Of course, most Members claimed that they had done so as to appear as though they cared about transparency in government, knowing full well that they would never be caught. But I had a plan: I found a few Republican House Members who were willing to take the risk with me to memorize one page of signatures while they signed themselves, and report those names back to me. I compiled the list of names, gave it to the *Wall Street Journal*, and let the chips fall where they may.

Of course, no Democrats had signed. I spent the rest of August recess on radio shows in each of the Democratic districts explaining my Discharge Petition, and I was pleased that there was such a groundswell of support from across America. The way the system worked is that once the

218th member signed, the signatures stopped. By the end of that August recess, the remaining Democratic members, under pressure from their constituents, were begging me to include their names, since they could not be added after the 218th name.

For sixty years, no one had dared to challenge this rule; even if they had wanted to, many were intimidated by the threat that they could be expelled. During that time, I was asked if I was worried about being thrown out of the House of Representatives. I replied that I would simply run in my own special election to replace me and then fill my own vacancy. Besides, there was no way the Democrats could expel me after that.

I was proud to learn later that this became known as one of the most significant reforms in the history of the House of Representatives and I was happy that my efforts were appreciated. As the *Daily Oklahoman*, wrote, "Inhofe's victory...is one of the most significant changes in congressional rules in the modern era."[33]

POISED TO TAKE THE GAVEL

When I was elected to the U.S. Senate, I knew from the start exactly what committees I wanted to serve on: Armed Services, and Environment and Public Works. My reason for wanting to be on Armed Services was simple. I am proud to have served in the U.S. Army. It was probably the best thing that could have happened to me, as I developed a profound appreciation for the price of freedom.

When I was elected to the Oklahoma Legislature in 1967, one of my first duties was to travel to Washington to appear before the Environment and Public Works Committee, when Senator Jennings Randolph was the Chairman, to protest Ladybird's Highway Beautification Act of 1965. Little did I know then that this would be the Committee I would ultimately end up chairing.

Since my goal was to do whatever I could to rein in the kinds of regulations that stifle entrepreneurship and job growth, I felt it was important to serve on the Environment and Public Works Committee. This committee has primary jurisdiction over the Environmental Protection Agency—an agency that puts forth some of the most job-destroying regulations in the country. From water rules, to the regulation of our energy capabilities, the EPA has the power to affect almost every facet of business and industry in America, and under the Obama administration, the results have been devastating for our economy.

These overreaching regulations especially have huge impacts on my state of Oklahoma, which is one of the top energy producing states in the country. I'm a pilot and I often fly my plane around the state to different political and civic events. When I fly into Will Rogers Airport in Oklahoma City and look out the window, I'm always greeted by multiple pump jacks located just off the runways, and there are more still pumping alongside the highways as you travel through the state. It's second nature to see that in Oklahoma—we're known for our energy development, so much so that old "Petunia #1," an oil well named for being in the middle of a flower bed, stands tall in front of the Oklahoma State Capitol building. It no longer produces oil, but it's a reminder of past and present Oklahoma.

Oklahoma has been blessed with tremendous resources, and developing those resources has led to a huge economic boost and the creation of good paying jobs.[34] In fact, Oklahoma's unemployment rate is a far cry from the national unemployment rate. This good state economy is in large part due to our strong energy development industry. The U.S. Energy Information Administration reports that presently Oklahoma has over 83,000 producing oil wells and well over 43,000 producing natural gas wells.[35] Oklahoma City University published a study in 2009, which found that the oil and gas industry in Oklahoma is responsible for 300,000 jobs in the state and contributes $51 billion to the state economy in just one year alone. In fact, twenty-percent of state revenues are due to the oil and gas industry.[36] In addition to oil and gas, Oklahoma is also one of the largest wind energy-generating states. I have always said that the best way to power this machine called America is through an all-of-the above energy policy that includes fossil fuels and renewables—and Oklahoma provides a good example of that.

Far from being a friend of Big Oil, as the accusation is often levied against me, I am a friend of "Little Oil" or of any Oklahomans who strive to develop our vast resources. Most of those involved in energy development in Oklahoma are running small businesses, like A.W. Swift. They are the people I fight for every day in Washington.

Many environmentalists see me as their enemy because they measure the "greenness" of politicians by how many federal laws they impose on the American people. In contrast, I have always believed that the environment is best served when the economy is strong, and we *can* develop our resources while being good stewards of the environment.

On August 3, 2007, Emily Belz of *The Hill* contacted me to ask if I consider myself a green lawmaker.[37] I said absolutely: I have always strived to be a good steward of the environment and I see myself as a conservationist. One of my favorite stories involves Ila Loetscher, a remarkable person who is known to most as the "Turtle Lady." I met Ila over forty years ago while spending time on South Padre Island with my family. She used to rescue turtles from nets and traps—some of them had torn flippers and cracked shells. She trained her turtles to clap their flippers and roll over in the water on cue. Our kids loved to watch them perform. The Ridley Sea Turtles lay their eggs, cover them, and leave. When the eggs hatch, the tiny turtles must struggle on their own toward the water, and those who make it, swim away. It wasn't long before Ila had all of us out on the beaches of South Padre late at night guarding and protecting newly hatched Ridley Sea Turtles as they made their first journey from the beach to the ocean.

As part of being a conservationist, I have always believed that personal responsibility breeds environmental stewardship. I saw this so often when I was Mayor of Tulsa: whenever individuals were involved in efforts to protect the species and the environment, the outcome is always more effective and efficient than it would be with regulations solely from the federal government. One of the best examples of this is the Partners for Fish and Wildlife Program, which was authorized when I was Chairman of the Environment and Public Works Committee. Whereas regulations under the Endangered Species Act have a low success rate in recovering species but are highly successful in stifling economic growth, the Partners Program is much more effective in preserving the species and the economy because it works with property owners instead of against them.

Good energy and environment policy is about achieving a healthy balance between environmental progress and economic growth. The global warming regulations promoted by President Obama, whether through cap and trade or through regulations by the EPA, completely lack that balance, so much so that they would destroy our economy while doing nothing to help the environment. Looking back, it is clear that the global warming debate was never really about saving the world; it was about controlling the lives of every American. MIT climate scientist Richard Lindzen summed up it up perfectly in March 2007 when he said, "Controlling carbon is a bureaucrat's dream. If you control carbon, you control life."[38]

Having every aspect of one's life controlled is completely against everything America stands for, and that's why global warming is one of the defining debates of our era. These experiences in my life, from reining in job-killing regulations as Mayor of Tulsa, to being the voice in the wilderness on the balanced budget amendment, to my one-man truth squad battle in the House, even to rescuing turtles in South Padre island are the training grounds that prepared me to fight this huge battle: The Greatest Hoax.

2

"THE MOST DANGEROUS MAN ON THE PLANET": EXPOSING THE HOAX

W HEN I ARRIVED IN MILAN, Italy, in December 2003 for the annual IPCC global warming conference, I was in for a big surprise. While I was fully aware that I would be walking straight into the lion's den, I certainly wasn't expecting to find my face on a WANTED poster for being "The most dangerous man on the planet." My crime was clearly laid out; under a picture of me at the dais was the quote that had made me famous with the environmental crowd just a few months before: "Global warming is the greatest hoax ever perpetrated on the American people."

I looked around and realized that the posters were plastered everywhere throughout the convention center, and there were some taped to telephone poles on the city streets. When I first discovered that I was the most dangerous man on the planet, I must admit, I was a little stunned. But then I thought of how my family and friends would smile when they heard the news and I knew my grandchildren would be impressed. Up until that point, the closest I had ever come to being called "dangerous" was when I was repeatedly referred to as the "Renegade Conservative from Oklahoma"—a title I have long been proud to have. Now because I had dared to question the validity of the radical global warming agenda, I was suddenly the world's most wanted climate criminal. With so much wrath directed at me, my colleagues even suggested that I may be a target in Italy and perhaps I should request that the conference provide me with additional security.

But if they were trying to intimidate me, it didn't work. My staff tried to stop me as I made my way over the room of the National Environmental Trust, the group responsible for the incriminating posters. When they saw me coming they looked worried—they thought I was going to be mad, but instead I smiled and shook their hands. I told them, "I'm just glad you guys got my quote right this time. You know, you usually misquote me."

The young man behind the table asked me to autograph one of the posters and I said I'd be honored. "Great to have friends like you, Jim Inhofe," I wrote. It turns out that my autographed poster hung framed in their Washington, D.C., office for years. It may still be there.

TAKING UP THE GAVEL

The WANTED posters incident was a critical juncture when I came to understand firsthand just how far the environmental left will go to isolate and silence anyone who dares to call their agenda into question. Instead of engaging in an intellectual debate about the problems I had exposed in the scientific process underlying their theories, they resorted to threats and attacks. And that is precisely why they ultimately lost the debate. I've always said that when you don't have science on your side, when you don't have logic on your side, when you don't have truth on your side, you resort to attacks.

For a long time, it worked: once intimidated, many of my colleagues would either change their position on global warming, or they would stay quiet so as not to be a target. In fact, the green movement had a clear record of success in silencing dissenters—that is, until I became Chairman of the Environment and Public Works Committee in 2003. Not only did I consistently speak out myself, I made it a priority to ensure that other silenced voices, especially in the scientific community, were heard as well.

From the moment the new committee leadership was in place, I think the Environment and Public Works Committee didn't know what hit them. I could not have been more different from my predecessors. In fact, as a staunch conservative from Oklahoma, an energy-developing state, I was a radical departure from my colleagues who had held my position before, including Senators Jim Jeffords of Vermont, Robert Smith of New Hampshire, and John Chafee of Rhode Island, who were, for the most part, less conservative and from eastern states with much different constituencies. More often than not, they believed that the more regulations from Washington, the better for our environment and nation. I never saw it that way; I have always believed that we need to achieve a healthy balance between environmental progress and economic growth. In fact, I was the first Chairman to invite witnesses from industry and energy development sectors to testify on how excessive environmental regulations may affect their ability to create jobs or expand their businesses. Also for the first recorded time in the Committee's history, I made sure that it was staffed with people who had actually worked for a living, instead of filling it with the kinds of idealists who often end up in these jobs.

Because the Environment and Public Works Committee has primary jurisdiction over the issue of global warming, I realized that as Chairman, I had a profound responsibility, as any "solution" to global warming would have far-reaching impacts for our nation. That's why from the moment I took up the gavel, I established three key principles for our work on the committee: (1) it should rely on the most objective science, (2) it should consider costs on businesses and consumers, and (3) the bureaucracy should serve, not rule, the people. I knew that without these principles, we could not make effective public policy decisions. These three principles would guide us as we continued to improve the environment, while also encouraging economic growth and prosperity. It was a fundamental shift in the way the Committee operated. For the first time in environmental policy, instead of looking at a problem with the mindset of "How can the federal government solve the problem?" we looked at "What problems were the federal government causing?"

KYOTO TREATY ALL ECONOMIC PAIN FOR NO ENVIRONMENTAL GAIN

In the 1990s, as the global warming hysteria was heating up, the so-called solution was the Kyoto Protocol, a treaty that required nations that were signatories to reduce their greenhouse gas emissions by considerable amounts below 1990 levels; specifically, the U.S. would have to reduce its emissions 31 percent below the level otherwise predicted for 2010. To put this in perspective, as the Business Roundtable pointed out, that target was "the equivalent of having to eliminate all current emissions from either the U.S. transportation sector, or the utilities sector [residential and commercial sources], or industry."[39]

The Clinton Administration, led by then Vice President Al Gore, signed Kyoto on November 12, 1998, but never submitted it to the Senate for ratification. That's because the Senate sent a powerful message to President Clinton. By a vote of 95-0, the Senate passed the Byrd-Hagel resolution on July 25, 1997, which stated that the Senate would not ratify a treaty if (1) it caused substantial economic harm and (2) if developing countries were not required to participate on the same timetable.[40]

Of course, Kyoto satisfied neither of the requirements of the Byrd-Hagel resolution. One definitive study from Wharton Econometric Forecasting Associates, or WEFA, a private consulting company founded by professors from the University of Pennsylvania's Wharton Business School, revealed that Kyoto would cost 2.4 million U.S. jobs and reduce GDP by 3.2 percent.[41] In other words, the pricetag for the United States would be over $300 billion annually. It found that Americans would face higher food, medical, and housing costs—for food, an increase of 9 percent; medicine, an increase of 11 percent; and housing, an increase of 21 percent. At the same time, an average household of four would see its real income drop by $2,700 in 2010, and each year thereafter. Under Kyoto, energy and electricity prices would nearly double, and gasoline prices would go up an additional 65 cents per gallon. It was truly a recipe for economic disaster.

In July 2003, the Congressional Budget Office (CBO) found that "The price increases resulting from a carbon cap would be regressive— that is, they would place a relatively greater burden on lower-income households than on higher-income ones."[42] So it would have been a raw deal for America and a disaster for the poor, who would have to pay a disproportionate amount of their incomes on higher energy prices. I have always found it ironic that the environmental left continually claims the high moral ground, saying that their policies are to protect the most vulnerable, yet the very policies they espouse would cause the greatest harm to the poorest among us.

One witness we called before the Committee, Tom Mullen of the Cleveland Catholic Charities, put it the most succinctly:

> The elderly on fixed limited incomes and the working poor with families have made it clear to me on a daily basis that they cannot afford increases in costs for their basic needs. In Cleveland, over one-fourth of all children live in poverty and are in a family of a single female head of household. These children will suffer further loss of basic needs as their moms are forced to make choices of whether to pay the rent or live in a shelter; pay the heating bill or see their child freeze; buy food or risk the availability of a hunger center. These are not choices any senior citizen, child, or, for that matter, person in America should make.[43]

In 2003, those Americans who made less than $30,000 spent 22 percent of their take-home pay on energy costs, such as gasoline or their monthly utility bills. People who made less than $10,000 per year spent 68 percent of their take-home pay on energy. Many of those people live in my home state of Oklahoma and other rural and urban areas of the country that the East and West coast liberal elites refer to as flyover country.

One would think that for all the economic suffering that the United States would have to endure, surely it would be worth it; surely it would save the world. Not exactly. The Kyoto Treaty did not bind developing countries like China, India, Brazil, and Mexico, who were all signatories to Kyoto and some of the world's largest emitters of greenhouse gases, to any emission requirements. It's not difficult to predict the results: as jobs went overseas and to countries that do not have emissions limits, worldwide emissions would actually increase.

In the end, no matter how the environmental left tried to couch it; no matter what the politicians said to make their case, Kyoto would have been all economic pain for no environmental gain. Sometimes I feel like a broken record, as I have been repeating that message ever since I began this fight. In 2003, I told world leaders in Milan that the U.S. would never ratify Kyoto. Six years later, in Copenhagen in 2009, I told world leaders that the United States Senate would never pass cap and trade. They didn't like my message then and they don't like it now. But I was right and they were wrong.

KYOTO: NOT ABOUT SAVING THE PLANET

In truth, Kyoto's objective had nothing to do with saving the globe, because it was clear that was the one thing it would fail to do—it was purely political.

Before and during the time I was Chairman, the push for the United States to ratify Kyoto continued. In June 2001, Germany released a statement declaring that the world needs Kyoto because its greenhouse gas reduction targets "are indispensable."[44] Also that June, Swedish Prime Minister Goeran Persson flatly stated that Kyoto is necessary.[45] I couldn't help but ask: necessary for what? We already knew that it would not reduce emissions. In fact, at that time, according to the EU, Environment Ministry, most EU member states were not on schedule to meet their Kyoto targets.

Even from the beginning it was very clear that Kyoto was not about reducing emissions, but something much more sinister. One indication of this came from Russia, who ratified Kyoto not because the government believed in catastrophic global warming, but because ratification was Russia's key to joining the World Trade Organization. Also, under Kyoto, Russia could profit from selling emissions credits to the EU and continue business as usual, without undertaking economically harmful emissions reductions. They could make billions of dollars by *not* developing their own natural resources.

This made a big impression on me because I had the opportunity to fly an airplane around the world. In fact, I'm one of the relatively few private-plane pilots who has followed in the tradition of Oklahoman

Wiley Post, who traveled around the world in a small private plane in 1931. There's an amazing thing about flying in a small aircraft. You see the world from a different vantage point—closer to the ground than the major airliners but high enough to get a distinct perspective on the ground below, the sky above, and the horizon in the distance. I remember taking off from Moscow and flying over the vast wilderness of Siberia where they have all these natural resources. I was flying over time zone after time zone without seeing any sign of life—nothing but one east-west river that runs through there. All that time I was thinking, these engines are running pretty rough and its lousy gasoline they've got over here. What will I do if I end up down there? Just being up there and seeing for myself the vast land that contained a wealth of resources that would remain untapped—it seemed like such a waste.

But that's the way the system works: countries are rewarded for not developing their resources. While in Milan, I met with members of the Russian delegation, who told me, "of course we will sign the Kyoto Treaty, we don't believe in global warming, but we get so much funding from the UN and developed countries that we would be foolish not to sign. Also, the Treaty will be long dead before Russia would have to comply."

It was clear that the point of Kyoto wasn't to reduce emissions, so what was it? French President Jacques Chirac provided a good answer to that question at The Hague in November 2000 when he explained that Kyoto represents "the first component of an authentic global governance."[46] Margot Wallstrom, the EU's Environment Commissioner, took a slightly different view, but one that reveals the real motives of Kyoto's supporters. She asserted that Kyoto is about "the economy, about leveling the playing field for big businesses worldwide."[47] The meaning behind Chirac and Wallstrom's comments was clear to me: (1) Kyoto represented an attempt by certain elements within the international community to restrain U.S. interests; and (2) Kyoto was an economic weapon designed to undermine the global competitiveness and economic superiority of the United States. Canadian Prime Minister Stephen Harper put it well when he later called Kyoto a "socialist scheme."[48] In short, Kyoto went against everything the United States stands for.

THE GREATEST HOAX

When I became Chairman, I said that if the United States was even going to consider taking drastic measures on global warming, the science behind those decisions had better be absolutely sound.

I am most remembered for standing on the Senate floor in July 2003 and declaring that man-made catastrophic global warming is the greatest hoax ever perpetrated on the American people. I am often accused of coming to that conclusion flippantly—as many on the environmental left have said, just because I didn't like what I heard from the mainstream global warming alarmists. Nothing could be farther from the truth. I came to that conclusion only after engaging in a lengthy, rigorous oversight process over the course of a few years; it was the most thorough investigation of the science by any senator. In more than twelve thousand words and several hours on the floor over the course of two days, I brought numerous inconsistencies and gaps in the mainstream theory to light. Only after that did I conclude that the science to justify the catastrophic theory simply was not there.

My painstaking oversight was almost to a fault, so much so that one editorial in the *Oklahoman*, looking back on a number of my science speeches, said that my "detailed and highly technical fodder" might "have a dual use as a cure for insomnia." Although I may have been putting everyone to sleep with my exhaustive approach, I was also appreciative that the editorial recognized the importance of my work. As it also said, "Credit Inhofe for nimbly making his case. And we think he's got a point. The science on human causation of global warming is conflicted and unsettled. There's something to be said for a senator who does his homework and is willing to swim against the stream on this important issue."[49]

My "Greatest Hoax" speech was the first speech I gave on the Senate floor as Chairman of the Environment and Public Works Committee, and I knew that it was going to be a defining moment for me going forward in this leadership role. The night before I gave the speech, I spent hours going through it with one of my most trusted aides. At the end of our meeting, I told him how I wanted to close: with my famous line. I took out a pen and handwrote it at the conclusion of the speech, "With all of the hysteria, all of the fear, all of the phony science, could it be that man-made global warming is the greatest hoax ever perpetrated on the American people? It sure sounds like it."

He looked slightly pained and asked seriously, "Are you sure?" He asked me that question not because there was any doubt that it was the message I needed to convey. He was asking me if I was prepared to endure the wrath of the environmental left for that comment, for the most part all alone—which clearly manifested itself months later in Milan. He tried to talk me out of it but at that moment, I couldn't have been more sure; catastrophic global warming was the greatest hoax, and it was my responsibility to expose it for what it was. The next day I asked him, "Now you made sure to put that line in the speech, right?"

Hopefully without offering my readers a cure for insomnia, it is important to explain how I came to that conclusion.

CATASTROPHIC GLOBAL WARMING BASED ON FEAR, NOT SCIENCE

I am a U.S. Senator, and a former mayor and businessman. I am not a scientist. But I do understand politics. And the more I delved into the science purporting global catastrophe, the more I saw the extent to which the science was being co-opted by those who care more about peddling fear of gloom and doom to further their own, broader political agendas.

As I said on the Senate floor in July 2003, much of the debate about global warming is predicated on fear rather than science. The alarmists predicted a future plagued by catastrophic flooding, war, terrorism, economic dislocations, droughts, crop failures, mosquito-borne diseases, and harsh weather, and they placed the blamed squarely on man-made carbon emissions.

One thing that people don't always realize is that global warming alarmism became like a religion that divided the world into believers and skeptics—the latter of which was the most exclusive club in Washington, as it consisted primarily of me. At any rate, being a skeptic was to them heresy of the highest order. That, of course, is what landed my face on those WANTED posters for being the most dangerous man on the planet. I still find it amazing that, in their minds, I was capable of singlehandedly bringing about the end of the world. But such was their logic.

Of course, alarmists accused me of attacking the science of global warming—that is part of their game. But the truth is that throughout my battle against the hoax, I consistently defended credible, objective science by exposing the corrupting influences that continually subverted it for political purposes. Good policy must be based on good science, not on religion, and that requires science free from bias, whatever its conclusions.

GLOBAL WARMING OR GLOBAL COOLING?

I began my investigation by delving first into some of the most obvious inconsistencies of the catastrophe rhetoric.

My starting point was a quote from a particular *Newsweek* magazine article that said, "There are ominous signs that the Earth's weather patterns have begun to change dramatically and that these changes may portend a drastic decline in food production—with serious political implications for just about every nation on Earth."[50] Another came from *Time* magazine: "As they review the bizarre and unpredictable weather pattern of the past several years, a growing number of scientists are beginning to suspect that many seemingly contradictory meteorological fluctuations are actually part of a global climatic upheaval."[51] All of this climate rhetoric sounds very ominous. That is, until you realize that these passages come from articles in a 1975 edition of *Newsweek* magazine and *Time* magazine in 1974. These articles weren't referring to global warming; *they were warning of a coming ice age.*

These fears can also be found in a 1974 study by the National Science Board, the governing body of the National Science Foundation, which stated, "During the last 20 to 30 years, world temperature has fallen, irregularly at first but more sharply over the last decade."[52] Two years earlier, the board had observed: "Judging from the record of the past interglacial ages, the present time of high temperatures should be drawing to an end...leading into the next glacial age."[53]

Yet, not long after we went through a widespread global cooling scare, alarmists boldly went forward to assert that the science behind the phenomenon of global warming was "settled," the "debate was over," and that there was no question that it was man-made and catastrophic.

In truth, since 1895, alarmists have alternated between global cooling

SCIENCE

The Cooling World

There are ominous signs that the earth's weather patterns have begun to change dramatically and that these changes may portend a drastic decline in food production—with serious political implications for just about every nation on earth. The drop in food output could begin quite soon, perhaps only ten years from now. The regions destined to feel its impact are the great wheat-producing lands of Canada and the U.S.S.R. in the north, along with a number of marginally self-sufficient tropical areas—parts of India, Pakistan, Bangladesh, Indochina and Indonesia—where the growing season is dependent upon the rains brought by the monsoon.

The evidence in support of these predictions has now begun to accumulate so massively that meteorologists are hard-pressed to keep up with it. In England, farmers have seen their growing season decline by about two weeks since 1950, with a resultant over-all loss in grain production estimated at up to 100,000 tons annually. During the same time, the average temperature around the equator has risen by a fraction of a degree—a fraction that in some areas can mean drought and desolation. Last April, in the most devastating outbreak of tornadoes ever recorded, 148 twisters killed more than 300 people and caused half a billion dollars' worth of damage in thirteen U.S. states.

Trend: To scientists, these seemingly disparate incidents represent the advance signs of fundamental changes in the world's weather. The central fact is that after three quarters of a century of extraordinarily mild conditions, the earth's climate seems to be cooling down. Meteorologists disagree about the cause and extent of the cooling trend, as well as over its specific impact on local weather conditions. But they are almost unanimous in the view that the trend will reduce agricultural productivity for the rest of the century. If the climatic change is as profound as some of the pessimists fear, the resulting famines could be catastrophic. "A major climatic change would force economic and social adjustments on a worldwide scale," warns a recent report by the National Academy of Sciences, "because the global patterns of food production and population that have evolved are implicitly dependent on the climate of the present century."

A survey completed last year by Dr. Murray Mitchell of the National Oceanic and Atmospheric Administration reveals a drop of half a degree in average ground temperatures between 1945 and 1968. According to George Kukla of Columbia University, satellite photos indicated a sudden, large increase in Northern Hemisphere snow cover in the winter of 1971–72. And

a study released last month by two NOAA scientists notes that the amount of sunshine reaching the ground in the continental U.S. diminished by 1.3 per cent between 1964 and 1972.

To the layman, the relatively small changes in temperature and sunshine can be highly misleading. Reid Bryson of the University of Wisconsin points out that the earth's average temperature during the great Ice Ages was only about 7 degrees lower than during its warmest eras—and that the present decline has taken the planet about a sixth of the way toward the Ice Age average. Others regard the cooling as a reversion to the "little ice age" conditions that brought bitter winters to much of Europe and northern America between 1600 and 1900—years when the Thames used to freeze so solidly that Londoners roasted oxen on the ice and when iceboats sailed the Hudson River almost as far south as New York City.

Just what causes the onset of major and minor ice ages remains a mystery. "Our knowledge of the mechanisms of climat-

ic change is at least as fragmentary as our data," concedes the National Academy of Sciences report. "Not only are the basic scientific questions largely unanswered, but in many cases we do not yet know enough to pose the key questions."

Extremes: Meteorologists think that they can forecast the short-term results of the return to the norm of the last century. They begin by noting the slight drop in over-all temperature that produces large numbers of pressure centers in the upper atmosphere. These break up the smooth flow of westerly winds over temperate areas. The stagnant air produced in this way causes an increase in extremes of local weather such as droughts, floods, extended dry spells, long freezes, delayed monsoons and even local temperature increases—all of which have a direct impact on food supplies.

"The world's food-producing system," warns Dr. James D. McQuigg of NOAA's Center for Climatic and Environmental Assessment, "is much more sensitive to the weather variable than it was even five years ago." Furthermore, the growth of world population and creation of new national boundaries make it impossible for starving peoples to migrate from their devastated fields, as they did during past famines.

Climatologists are pessimistic that political leaders will take any positive action to compensate for the climatic change, or even to allay its effects. They concede that some of the more spectacular solutions proposed, such as melting the arctic ice cap by covering it with black soot or diverting arctic rivers, might create problems far greater than those they solve. But the scientists see few signs that government leaders anywhere are even prepared to take the simple measures of stockpiling food or of introducing the variables of climatic uncertainty into economic projections of future food supplies. The longer the planners delay, the more difficult will they find it to cope with climatic change once the results become grim reality.

—PETER GWYNNE with bureau reports

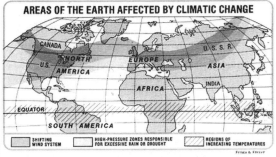

AREAS OF THE EARTH AFFECTED BY CLIMATIC CHANGE

SHIFTING WIND SYSTEM — HIGH-PRESSURE ZONES RESPONSIBLE FOR EXCESSIVE RAIN OR DROUGHT — REGIONS OF INCREASING TEMPERATURES

AVERAGE TEMPERATURE CHANGE
Source: National Center for Atmospheric Research

64

and warming scares during four separate and sometimes overlapping time periods. From 1895 until the 1930s, the media peddled a coming ice age—and the world was coming to an end. From the late 1920s until the 1960s, they warned of global warming—and again the world was coming to an end. From the 1950s until the 1970s, they warned us again of a coming ice age—and as before, the world was coming to an end. The latest apocalyptic scare about global warming is the fourth attempt to promote catastrophe and the world has yet to come to an end.

While alarmists continued their message of climate apocalypse, I maintained that it was very simplistic to say that a one degree Fahrenheit temperature increase during the twentieth century means that we are all doomed. After all, a one degree Fahrenheit rise has coincided with the greatest advancement of living standards, life expectancy, food production and human health in the history of our planet. So it is hard to argue that the global warming we experienced in the twentieth century was somehow negative or part of a catastrophic trend.

Of course, these particular inconsistencies in the alarmists' theories were the kind of observations that one could point out without delving very far into the debate. As I continued my investigation, more and more serious inconsistencies with the science came to light.

THE UNITED NATIONS AND THE EMERGENCE OF THE IPCC

Even though the global community has been somewhat critical of the United States in recent years, our nation remains one of the most generous in the world. The American people have always gone above and beyond the call of duty to help our neighbors and far away friends who are in need.

You probably remember the aftermath of 2010's devastating earthquake in Haiti. According to one report published by the Inter-American Development Bank, the disaster killed over 200,000 people and could require more than $7 billion in recovery costs.[54] How did Americans respond? With an outpouring of generosity and support. According to the *Wall Street Journal*, Americans "raised more than $150 million in four days for the Haiti relief efforts."[55] Countless others traveled to Haiti to help with the cleanup and rescue efforts. This is not an uncommon headline following natural disasters. The same response occurred after the tsunamis in Japan and the Indian Ocean. The American people are generous, and I believe that responses like these are a natural display of our nation's character.

This is important when considering the global warming issue because if the American people truly believed that the effects of climate change were man-made and going to cause the destruction of the earth, then I believe the American people would be the first to stand up and provide a solution. But this has not happened.

While the global warming issue did not come to the attention of the American people until the late 1980s or early 1990s, the seeds of the hoax were being sown decades before by environmental elites at the United Nations.

Why would they want to do this? I believe it is because many who work at the United Nations would like to see the institution's mandate expanded well beyond its original intent. I do not think they are satisfied with having an influential role in the international arena—I think they want a controlling one. They want the United Nations to have sovereignty, control of the world's economic and political systems, and the ability to tax wealthy nations and redistribute their resources to poorer ones in an effort, as Margo Wallstrom said, to "level the playing field."[56]

Just like liberals in Washington, the elites at the United Nations truly believe that they know how to run things better than any individual country ever could. In this way, they are like "super-liberals" on an international scale. On one of its websites, the UN even claims that its "moral authority" is one of its "best properties."[57] But what do the elites and super-liberals at the UN believe?

Briefly, their guiding philosophy is known as "sustainable development," and it is alarmingly similar to the utopian ideals of global socialism. Unfortunately for us Americans, this philosophy would require those of us in Western nations to surrender our lifestyles and our resources so that they could be redistributed to developing nations, just as our friends from Russia acknowledged. And who would be doing this redistribution? You got it—the United Nations.

Perhaps what is most alarming is that this is all being done in an effort to save the "environment," which is why I believe the leaders at the UN have been pushing the global warming issue so hard. In fact, the Kyoto Protocol embodies many of the actual goals and priorities of the UN super elites. That is why it's a crucial component of their effort to make the UN more powerful and important.

More details about the history of sustainable development, including how the philosophy was developed and what is wrong with it, are included in Appendix A. I encourage you to check that out and see all that the UN is up to. Keep these ideas in mind as you read the rest of the book. If you do, you'll be able to see how the global warming issue fits in with the bigger picture of the ambitions of the super-liberals at the UN.

But for now, let's stay focused on the core of the story. The UN began officially working on the global warming issue when it decided to create the IPCC.

IPCC SCIENTIFIC INTEGRITY QUESTIONED

Like most bad things that come to America, the primary science behind catastrophic global warming came from the United Nations—specifically from its Intergovernmental Panel on Climate Change (IPCC). Again, a careful reading of Appendix A is a must to understand thoroughly the origin of the hoax.

The Kyoto Treaty explicitly acknowledged that man-made emissions, primarily from fossil fuel development, are causing global temperatures to rise to catastrophic levels. Those who sign on to Kyoto pledge to cut back dramatically or even work to eliminate fossil fuels so that the world can return to global temperatures at "normal" levels. According to the IPCC, Kyoto was to achieve "stabilization of greenhouse gas concentrations in the atmosphere at a level that would prevent dangerous anthropogenic interference with the climate system."[58]

But when it came to discovering what those "normal" levels were, the IPCC couldn't provide a scientific explanation. That's because they didn't have one. Dr. S. Fred Singer, formerly an atmospheric scientist at the University of Virginia, said, "No one knows what constitutes a 'dangerous' concentration. There exists, as yet, no scientific basis for defining such a concentration, or even of knowing whether it is more or less than current levels of carbon dioxide."[59] This was seriously problematic. My question was: how can we bring greenhouse gas concentrations to normal levels if we don't know what those normal levels are?

The more questions I asked, the clearer it became that Kyoto emissions reduction targets were arbitrary, lacking in scientific basis. This was not just my opinion, but the conclusion reached by the country's most recognized climate scientists. Dr. Tom Wigley, one of Al Gore's own scientists, was one of them. After President Clinton signed on to the Kyoto Protocol in 1997, Dr. Wigley, a senior scientist at the National Center for Atmospheric Research, found that if the Kyoto Protocol were fully implemented by all signatories, it would reduce temperatures by a mere 0.06 degrees Celsius by 2050.[60] And that's if the United States had ratified Kyoto and the other signatories met their targets. What does this mean? Such an amount is so small that ground-based thermometers cannot reliably measure it.

Dr. Richard Lindzen, an MIT scientist and member of the National Academy of Sciences who specializes in climate science, told the Environment and Public Works Committee on May 2, 2001, that there is a "definitive disconnect between Kyoto and science. Should a catastrophic scenario prove correct, Kyoto would not prevent it."[61] Similarly, Dr. James Hansen of NASA, the father of global warming theory citing Wigley and Malakoff, said that Kyoto Protocol "will have little effect" on global temperature in the twenty-first century. In a rather stunning follow-up, Hansen said it would take thirty Kyotos to reduce warming to an acceptable level.[62] If one Kyoto devastates the American economy, what would thirty do? So even the scientists were saying that it would be all pain for no gain.

In December 2004, when several nations including the United States met in Buenos Aires for the tenth round of international climate change negotiations, I was happy that the U.S. delegation held firm both in its categorical rejection of Kyoto and the questionable science behind it. Paula Dobriansky, Undersecretary of State for Global Affairs, and the leader of the U.S. delegation, put it well when she told the conference, "Science tells us that we cannot say with any certainty what constitutes a dangerous level of warming, and therefore what level must be avoided."[63]

FLAWED IPCC ASSESSMENT REPORTS

Our rigorous oversight of the IPCC began with my "Greatest Hoax" speech in 2003 and continued over the course of many years. In numerous speeches, I recounted the systematic and documented abuse of the scientific process by the IPCC, which claims it provides the most complete and objective scientific assessment in the world on the subject of climate change.

In 2003, I began to expose the flaws in the IPCC process that were glaringly apparent in the first IPCC Assessment Report in 1990, which found that the climate record of the past century was "broadly consistent" with the changes in Earth's surface temperature, as calculated by climate models that incorporated the observed increase in greenhouse gases. This conclusion, however, appeared suspect to me, considering the climate cooled between 1940 and 1975, just as industrial activity grew rapidly after World War II. How does one reconcile this cooling with the observed increase in greenhouse gases?

But the flaws revealed themselves in earnest when the IPCC issued its second assessment report in 1996. The most obvious problem was the altering of the document on the central question of whether man is causing global warming. Here is what Chapter 8—the key chapter in the report—stated on this central question in the final version accepted by reviewing scientists: "No study to date has positively attributed all or part [of the climate change observed] to anthropogenic causes."[64]

But when the final version was published, this and similar phrases in fifteen sections of the chapter were deleted or modified. Nearly all the changes removed hints of scientific doubts regarding the claim that human activities are having a major impact on global warming. In the *Summary for Policymakers*—which is the only part of the report that most reporters and policymakers read—a single phrase was inserted. It reads: "The balance of evidence suggests that there is a discernible human influence on global climate."[65]

The lead author for Chapter 8, Dr. Ben Santer, is not fully to blame for manipulation of the message. According to the journal *Nature*, the changes to the report were made in the midst of high-level pressure from the Clinton/Gore State Department to do so. In fact, after the State Department sent a letter to Sir John Houghton, Co-Chairman of the

IPCC, Houghton prevailed upon Santer to make the changes. Of course, the impact of this change was explosive, with media across the world, including heavyweights such as Peter Jennings, declaring this as proof that man is responsible for global warming. On September 10, 1995, the *New York Times* published an article titled "Global Warming Experts Call Human Role Likely." According to the *Times* account, the IPCC showed that global warming "is unlikely to be entirely due to natural causes."[66] When parsed, this account means fairly little. Not entirely due to natural causes? Well, how much, then? One percent? Twenty percent? Eighty-five percent?

The IPCC report was replete with caveats and qualifications, providing little evidence to support anthropogenic theories of global warming. The preceding paragraph in which the "balance of evidence" quote appears makes exactly that point. It reads, "Our ability to quantify the human influence on global climate is currently limited because the expected signal is still emerging from the noise of natural variability, and because there are uncertainties in key factors. These include the magnitude and patterns of long-term variability and the time evolving pattern of forcing by, and response to, changes in concentrations of greenhouse gases and aerosols, and land surface changes."

Perhaps one of the most important yet most ignored aspects of the IPCC report is that it is actually quite explicit about the uncertainties surrounding a link between human actions and global warming: "Although these global mean results suggest that there is some anthropogenic component in the observed temperature record, they cannot be considered compelling evidence of a clear cause-and-effect link between anthropogenic forces and changes in the Earth's surface temperature." Remember, the IPCC is supposed to provide the scientific basis for the alarmists' conclusions about global warming. Yet, even the IPCC admitted that its own science cannot be considered compelling evidence.

Dr. John Christy, professor of Atmospheric Science and director of the Earth System Science Center at the University of Alabama in Huntsville, and a key contributor to the 1995 IPCC report, participated with the lead authors in the drafting sessions and in the detailed review of the scientific text. He wrote in the *Montgomery Advertiser* on February 22, 1998, that much of what passes for common knowledge in the press regarding climate change is "inaccurate, incomplete or viewed out of context."[67] Dr. Christy said that many of the misconceptions about climate change originated from the IPCC's six-page executive summary, rather than the final report. It was the most widely read and quoted of the three documents published by the IPCC's Working Group, but, Christy said—and this point is crucial—it had the "least input from scientists and the greatest input from non-scientists."[68]

IPCC PLAYS HOCKEY AND LOSES

Five years later, the IPCC was back at it again, this time with the Third Assessment Report on Climate Change. In October 2000, the IPCC *Summary for Policymakers* was leaked to the media, which once again accepted the IPCC's conclusions as fact.

Based on the summary, *The Washington Post* wrote on October 30, 2000, "The consensus on global warming keeps strengthening."[69] In a similar vein, the *New York Times* confidently declared on October 28, 2000, "The international panel of climate scientists considered the most authoritative voice on global warming has now concluded that mankind's contribution to the problem is greater than originally believed."[70] Of course, these accounts were worded to maximize the fear factor, and upon closer inspection they had no compelling intellectual content. "Greater than originally believed"? Was that .01 percent, or 25 percent? And how much is greater? Double? Triple?

Such reporting prompted testimony by Dr. Richard Lindzen before the Committee on Environment and Public Works in May of 2001. Lindzen said, "Almost all reading and coverage of the IPCC is restricted to the highly publicized Summaries for Policymakers, which are written by representatives from governments, NGOs and business; the full reports, written by participating scientists, are largely ignored."[71] Of course, the Policymaker's Summary was politicized and radically differed from an earlier draft. For example, the draft concluded the following concerning the driving causes of climate change:

> From the body of evidence since IPCC (1996), we conclude that there has been a discernible human influence on global climate. Studies are beginning to separate the contributions to observed climate change attributable to individual external influences, both anthropogenic and natural. This work suggests that anthropogenic greenhouse gases are a substantial contributor to the observed warming, especially over the past 30 years. However, the accuracy of these estimates continues to be limited by uncertainties in estimates of internal variability, natural and anthropogenic forcing, and the climate response to external forcing.

The final version, however, looked quite different and concluded instead, "In the light of new evidence and taking into account the remaining uncertainties, most of the observed warming over the last 50 years is likely to have been due to the increase in greenhouse gas concentrations."

This kind of distortion was not unintentional. As Dr. Lindzen explained before the EPW Committee, "I personally witnessed coauthors forced to assert their 'green' credentials in defense of their statements."[72] In short, some parts of the IPCC process resembled a Soviet-style trial: facts were predetermined and ideological purity trumped technical and scientific rigor.

The most egregious flaw in the Third Assessment is undoubtedly the now infamous hockey stick graph produced by Dr. Michael Mann and others, which the IPCC enthusiastically embraced.

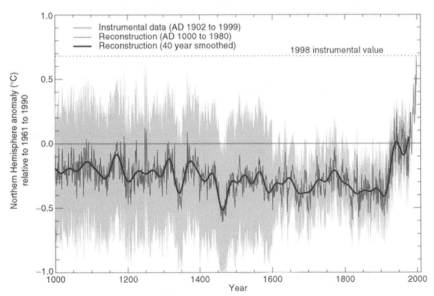

Intergovernmental Panel on Climate Change, Working Group 1, Climate Change Tool: The Scientific Basis, 2001

Mann's study concluded that the twentieth century was the warmest on record in the last one thousand years, showing flat temperatures until 1900, and then spiking upward. Put simply, it looked like a hockey stick. The cause for such a shift, of course, is attributed to industrialization and man-made greenhouse gas emissions. The conclusion is that industrialization, which spawned widespread use of fossil fuels, was causing the planet to warm.

The hockey stick achieved instant fame as proof that humans were causing global warming because it was featured prominently in the Summary Report read by the media—and eventually Al Gore made it mainstream in his documentary, *An Inconvenient Truth.*

But the problems with Mann's study were immense. First of all, it focuses on temperature trends only in the Northern Hemisphere. Mann extrapolated that data to reach the conclusion that global temperatures remained relatively stable and then dramatically increased at the beginning of the twentieth century. That leads to Mann's conclusion that the twentieth century has been the warmest in the last one thousand years. As is obvious, however, such an extrapolation cannot provide a reliable global perspective of long-term climate trends.

Mann's conclusions were also drawn mainly from twelve sets of climate proxy data, of which nine were tree rings, while the remaining three came from ice cores. Ice core data was drawn from Greenland and Peru. What's left is a picture of the Northern Hemisphere based on eight sets of tree-ring data. Again, hardly a convincing global picture of the last one thousand years.

In other words, Mann's hockey stick completely dismisses both the Medieval Warm Period (800 to 1300) and the Little Ice Age (1300 to 1900), two climate events that are widely recognized in the scientific literature. Mann said the twentieth century was "nominally the warmest" of the past millennium and that the decade of the 1990s was the warmest decade on record. The Medieval Warm Period and Little Ice Age are replaced by a largely benign and slightly cooling linear trend in climate until 1900. But as is clear from a close analysis of Mann's methods, the hockey stick is formed by crudely grafting the surface temperature record of the twentieth century onto a pre-1900 tree-ring record.

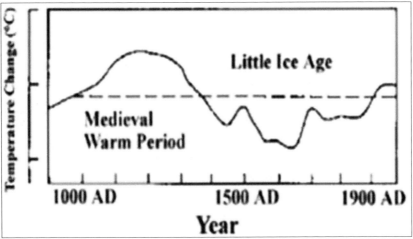

Intergovernmental Panel on Climate Change, Climate Change, The IPCC Scientific Assessment 202 (1990).

It was a highly controversial and scientifically flawed approach. As was widely recognized in the scientific community, two data series representing radically different variables (temperature and tree rings) cannot be grafted together credibly to create a single series. In simple terms, as Dr. Patrick Michaels of the University of Virginia explained, this is like "comparing apples to oranges."[73] Even Mann and his coauthors admitted that if the tree ring data set were removed from their climate reconstruction, the calibration and verification procedures they used would undermine their conclusions.

A study from Harvard-Smithsonian Center for Astrophysics strongly disputed Mann's methods and hypotheses. As coauthor Dr. David Legates wrote, "Although [Mann's work] is now widely used as proof of anthropogenic global warming, we've become concerned that such an analysis is in direct contradiction to most of the research and written histories available...Our paper shows this contradiction and argues that the results of Mann...are out of step with the preponderance of the evidence."[74] Indeed, Mann's theory of global warming was out of step with most scientific thinking on the subject. Dr. Hans von Storch, a prominent German researcher with the GKSS Institute for Coastal Research who believes in global warming put it this way: "We were able to show in a publication in *Science* that the [hockey stick] graph contains assumptions that are not permissible. Methodologically it is wrong: rubbish."[75]

Dr. von Storch was not the only one who felt that way. Three geophysicists from the University of Utah, in the April 7, 2004, edition of *Geophysical Research Letters*, concluded that Mann's methods used to create his temperature reconstruction were deeply flawed. In fact, their judgment was harsher than that. As they wrote, Mann's results are "based on using end points in computing changes in an oscillating series" and are "just bad science." I repeat: "just bad science."

My concerns about the "hockey stick" in those early days were validated later in 2005 when two Canadian researchers, Steven McIntyre and Ross McKitrick, essentially tore apart its statistical foundation. In essence, they discovered that Dr. Mann misused an established statistical method called principal components analysis (PCA). As they explained, Mann created a program that "effectively mines a data set for hockey stick patterns."[76] In other words, no matter what kind of data one uses, even if it is random and totally meaningless, the Mann method always produces a hockey stick. After conducting some 10,000 data simulations, the result was nearly always the same. "In over 99 percent of cases," McIntyre and McKitrick wrote, "it produced a hockey stick shaped PCI series."[77] Statistician Francis Zwiers of Environment Canada, a government agency, said he agreed that Dr. Mann's statistical method "preferentially produces hockey sticks when there are none in the data."[78] Even to a non-statistician, this looked extremely troubling. But that statistical error was just the beginning. On a public Web site where Dr. Mann filed data, McIntyre and McKitrick discovered an intriguing folder titled "BACKTO_1400-CENSORED." What McIntyre and McKitrick found in the folder was disturbing: Mann's hockey stick blade was based on a certain type of tree—a bristlecone pine—that, in effect, helped to manufacture the hockey stick.

So why was the bristlecone pine important? The bristlecone experienced a growth pulse in the Western United States in the late 19th and early 20th centuries. However, this growth pulse, as the specialist literature has confirmed, was not attributed to temperature. So using those pines, and only those pines as a proxy for temperature during this period is questionable at best. Even Mann's co-author has stated that the bristlecone growth pulse is a "mystery." Because of these obvious problems, McIntyre and McKitrick appropriately excluded the bristlecone data from their calculations. What did they find? Not the Mann hockey stick, to be sure, but a confirmation of the Medieval Warm

Period, which Mann's work had erased. As the CENSORED folder revealed, Mann and his colleagues never reported results obtained from calculations that excluded the bristlecone data. It appeared to be a case of selectively using data—that is, if you don't like the result, remove the offending data until you get the answer you want. As McIntyre and McKitrick explained, "Imagine the irony of this discovery…Mann accused us of selectively deleting North American proxy series. Now it appeared that he had results that were exactly the same as ours, stuffed away in a folder labeled CENSORED."

McIntyre and McKitrick believed there were additional errors in the Mann hockey stick. To confirm their suspicion, they need additional data from Dr. Mann, including the computer code he used to generate the graph. But Dr. Mann refused to supply it. As he told the *Wall Street Journal*, "Giving them the algorithm would be giving in to the intimidation tactics that these people are engaged in."[79]

Intimidation tactics? On April 12, 2005, I gave a speech on the Senate floor explaining McIntyre and McKitrick's findings and, as I said, "Mr. McIntyre and Mr. McKitrick were just trying to find the truth. What is Dr. Mann trying to hide?" Well, as the Climategate scandal eventually revealed, he may have been trying to "hide the decline" in temperatures, but we will get to that later.

In June 2006, the National Academy of Science released its study, "Surface Temperature Reconstructions for the Last 2,000 Years" which acknowledged that there were "relatively warm conditions centered around A.D. 1000 (identified by some as the 'Medieval Warm Period') and a relatively cold period (or 'Little Ice Age') centered around 1700." This report refuted the hockey stick which showed temperatures in the Northern Hemisphere remained relatively stable over nine hundred years, then spiked upward in the twentieth century. The NAS report also stated that "substantial uncertainties" surround Mann's claims that the last few decades of the twentieth century were the warmest in last one thousand years. In fact, while the report conceded that temperature data uncertainties increase going backward in time, it acknowledged that "not all individual proxy records indicate that the recent warmth is unprecedented." In one last blow, the NAS reports stated, "Even less confidence can be placed in the original conclusions by Mann et al. (1999) that 'the 1990s are likely the warmest decade and 1998 the warmest year in at least a millennium." When that report appeared, it only confirmed what I had been saying all along: the hockey stick was broken.

Climate alarmists have long been attempting to erase this inconvenient Medieval Warm Period from the Earth's climate history for at least a decade because it doesn't fit in with their theories of catastrophe. David Deming, who at the time was an Assistant Professor at the University of Oklahoma's College of Geosciences, understood this all too well. Dr. Deming was welcomed into the close-knit group of global warming believers after he published a paper in 1995 that noted some warming in the twentieth century. Deming says he was subsequently contacted by a prominent global warming alarmist who told him point-blank, "We have to get rid of that Medieval Warm Period."[80] When the "hockey stick" first appeared in 1998, it did just that. And yet, the IPCC immortalized the hockey stick as incontrovertible proof of catastrophic global warming.

IPCC: POLITICS, NOT SCIENCE

How could the IPCC so blatantly move forward with arguments based on such dubious science? I had an answer to that question as early as my 2003 speech.

First, the IPCC is a political institution. Its whole purpose is to support the efforts of the UN Framework Convention on Climate Change, which has the basic mission of eliminating the "threat" of global warming. This clearly creates a conflict of interest with the standard scientific goal of assessing scientific data in an objective manner. The *Summary for Policymakers* illustrates some of the problems with this: it was not approved by the scientists and economists who contributed to the report. It was approved by intergovernmental delegates—in short, politicians. It doesn't take a leap of imagination to realize that politicians will insist the report supports their political agenda. Both scientists and economists complained that the summary did not adequately reflect the uncertainties associated with tentative conclusions in the basic report. The uncertainties identified by contributing authors and reviewers seemed to disappear or were downplayed in the summary.

In other words, the lead authors and the chair of the IPCC control too much of the process. The old adage "power corrupts, and absolute power corrupts absolutely" applies. Only a handful of individuals were involved in changing the entire tone of the second assessment. Likewise, Michael Mann was a chapter lead author in the third assessment.

One stark example of how the process has been corrupted involves a U.S. government scientist who was among the world's most respected experts on hurricanes, Dr. Christopher Landsea, who eventually resigned as a contributing author of the fourth assessment. His reason was simple—the lead author for the chapter on extreme weather, Dr. Kevin Trenberth, had demonstrated he may pursue a political agenda linking global warming to more severe hurricanes. Trenberth had spoken at a forum where he was introduced as a lead author and proceeded forcefully to make the link. The only problem is that Trenberth's views may not have been widely accepted among the scientific community. As Landsea explained, "All previous and current research in the area of hurricane variability has shown no reliable, long-term trend up in the frequency or intensity of tropical cyclones, either in the Atlantic or any other basin."[81]

When Landsea brought it to the attention of the IPCC, he was told that Trenberth—who as lead author is supposed to bring a neutral, unbiased perspective to his position—would keep his position. Landsea concluded that, "because of Dr. Trenberth's pronouncements, the IPCC process on our assessment of these crucial extreme events in our climate system has been subverted and compromised, its neutrality lost."

Landsea's experience is not unique. Richard Lindzen, a prominent MIT researcher who was a contributing author to a chapter in the third assessment, said that the summary did not reflect the chapter to which he contributed. But when you examine how the IPCC is structured, is it really so surprising? The IPCC has consistently demonstrated an unreasoning resistance to accepting constructive critiques of its scientific and economic methods. I said that this was a recipe for de-legitimizing the entire endeavor in terms of providing credible information that is useful to policymakers.

I explained a few examples of this on the Senate floor in 2003. First, malaria is considered one of the four greatest risks associated with global warming. But the relationship between climate and mosquito populations is highly complex. There are more than thirty-five hundred species of mosquito, and all breed, feed, and behave differently. Yet the nine lead authors of the health section in the second assessment had published only six research papers on vector-borne diseases among them. Dr. Paul Reiter of the Pasteur Institute, a respected entomologist who has spent decades studying mosquito-borne malaria, believes that global warming would have little impact on the spread of malaria. But the IPCC refused to consider his views in its third assessment, and completely excluded him from contributing to the fourth assessment.

Here's another example: To predict future global warming, the IPCC estimated how much world economies would grow over the next century. Future increases in carbon dioxide emission estimates are directly tied to growth rates, which, in turn, drive the global warming predictions. However, the method the IPCC used to calculate growth rates was wrong. It also contained assumptions that developing nations will experience explosive growth—in some cases, becoming wealthier than the United States. These combine to greatly inflate even its lower-end estimates of future global warming. The IPCC, however, bowed to political pressure from the developing countries that refused to acknowledge the likelihood that they will not catch up to the developed world. The result: Future global warming predictions by the IPCC were based on a political choice, not on credible economic methodologies.

Likewise, the IPCC ignored the advice of economists who concluded that if global warming is real, future generations would have a higher quality of life if societies maximize economic growth and adapt to future warming rather than trying to drastically curb emissions. This problem with the economics led to a full-scale inquiry by the UK's House of Lords Select Committee on Economic Affairs, which found numerous problems with the IPCC. In fact, the problems they identified were so substantial, it led Lord Nigel Lawson, former Chancellor of the Exchequer and a Member of the Committee, to state: "I believe the IPCC process is so flawed, and the institution, it has to be said, so closed to reason, that it would be far better to thank it for the work it has done, close it down, and transfer all future international collaboration on the issue of climate change."[82]

These were the red flags that I examined with a vengeance from 2003 on. As more and more problems continued to surface with the IPCC, I warned that the entire institution would lose its credibility altogether if it did not take drastic steps to remedy the situation.

2005: REFORM NEEDED AT THE IPCC

In December 2005, as the IPCC was preparing to issue its fourth assessment report, I wrote to Dr. R.K. Pachauri, Chair of the Intergovernmental Panel on Climate Change, explaining my concern that certain scientific conclusions are selected or excluded from the IPCC's consideration and presentation, and how the science has been manipulated in order to reach a predetermined conclusion. I objected to the comments he made in Montreal earlier that year where he stated, "In the fourth assessment, we will conduct an extensive outreach effort. If facts are highlighted, not exaggerated...then it will help in changing public perception."[83] I told him that such an effort was in direct conflict with an objective assessment of the science, which should be free of political goals. Selective presentations of facts, whether accurate or not, skew the public's understanding of the issue by eliminating contrary findings and potentially considerable uncertainty about their accuracy.

I said that these problems must be remedied in order for the IPCC to present a fair and impartial conclusion as to the current state of climate science, and therefore regain its credibility. I wrote that the IPCC must adopt procedures that ensure that impartial scientific reviewers formally approve both the chapters *and* the *Summary for Policymakers*—the latter of which is the only document that members of the press and members of Congress ever read. When compared with the *actual* report, it was clear the *Summary for Policymakers* was being co-opted by activists with an agenda to shape the conclusions to show that man-made emissions were causing catastrophic global warming. To safeguard against the manipulation of the message, objective scientists, *not government delegates,* should be a part of the approval process. I also said that the IPCC must ensure that any uncertainties in the state of knowledge be clearly expressed in the *Summary for Policymakers*. But Dr. Pachauri dismissed my concerns. Here's how Reuters reported his response:

In the one-page letter, [Pachauri] denies that the IPCC has an alarmist bias and says, 'I have a deep commitment to the integrity and objectivity of the IPCC process.' Pachauri's main argument is that the IPCC comprises both scientists and more than 130 governments who approve IPCC reports line by line. That helps ensure fairness, he says.[84]

My concerns were confirmed when the IPCC process finally imploded in 2009 with the Climategate scandal and the errors in the IPCC science revealed in its wake, which showed once and for all that their process was all politics; science was secondary, even non-essential, to the ultimate goal of confirming catastrophic global warming and achieving global governance.

STATE OF FEAR: MICHAEL CRICHTON

As I was exposing flaws and inconsistencies of the IPCC science on the Senate floor, bestselling author Dr. Michael Crichton was doing it through fiction—it was, as the saying goes, art imitating life. In 2004, Dr. Crichton published *State of Fear*, a fascinating novel that questions the scientific consensus of man-made catastrophic global warming, while also predicting much of how the global warming debate would pan out. If I was the one-man truth squad on Capitol Hill, surely Dr. Crichton was the one-man truth squad in Hollywood.

With his scientific and medical background, Dr. Crichton made it a practice to do extensive research before writing his books. For *Congo*, he studied the Congo. In *Airframe,* he perfected the details of airframe structure. But my favorite bestseller of all was *State of Fear*. Initially, Dr. Crichton planned to write a novel about celebrities who warn about the dangers of global warming and the disasters that they predict come true. But the more he researched, the more he began to understand the true motives of the movement, and instead wrote a book about how it all turned out to be a scam. Of course, being from Hollywood and having been the force behind the blockbuster, *Jurassic Park*, as well as the very popular series *ER*, he knew the Hollywood elite mentality firsthand, and does an excellent job portraying young actors who go out to evangelize about global warming. In fact, one theme that *State of Fear* introduced and features brilliantly is the "religion" of global warming.

Throughout the book, environmental organizations are focused squarely on raising money, principally by scaring potential contributors with predictions of a global apocalypse and claims that they are saving the world.

What is truly remarkable about Dr. Crichton's book is just how accurately it establishes the progression of the entire movement. The novel calls out the media and Hollywood for its political agenda—which had not yet reached its height when he wrote the book in 2004. At one point in the story, a young actor called Ted Bradley films a scene where he lectures a group of third graders about the dangers of global warming. As he says:

But now these magnificent trees—having survived the threat of fire, the threat of logging, the threat of soil erosion, the threat of acid rain—now face their greatest threat ever. Global warming. You kids know what global warming is, don't you?

Hands went up all around the circle. "I know, I know!"

"I'm glad you do," Bradley said, gesturing for the kids to put their hands down. The only person talking today would be Ted Bradley. "But you may not know that global warming is going to cause a very sudden change in our climate. Maybe just a few months or years, and it will suddenly be much hotter or much colder. And there will be hordes of insects and diseases that will take down these wonderful trees."

[...]

By now the kids were fidgeting, and Bradley turned squarely to the cameras. He spoke with the easy authority he had mastered while playing the president for so many years on television. "The threat of abrupt climate change," he said, "is so devastating for mankind, and for all life on this planet, that conferences are being convened all around the world to deal with it. There is a conference in Los Angeles starting tomorrow, where scientists will discuss what we can do to mitigate this terrible threat. But if we do nothing, catastrophe looms. And these mighty, magnificent trees will be a memory, a postcard from the past, a snapshot of man's inhumanity to the natural world. We're responsible for catastrophic climate change. And only we can stop it."[85]

The novel also ventures into extensive detail about the science of the IPCC, demonstrating the manipulation of data from the very beginning. It even touches on some of the same problems I had exposed on the Senate floor in 2003. As Dr. Crichton writes,

> "...it is unquestionably manipulative. And Hansen's testimony wasn't the only instance of media manipulation that's occurred in the course of the global warming sales campaign. Don't forget the last-minute changes in the 1995 IPCC report."
>
> "IPCC? What last minute changes?"
>
> "The UN formed the Intergovernmental Panel on Climate Change in the late 1980s. That's the IPCC...a huge group of bureaucrats, and scientists under the thumb of bureaucrats. The idea was that since this was a global problem, the UN would track climate research and issue reports every few years. The first assessment report in 1990 said it would be very difficult to detect a human influence on climate, although everybody was concerned that one might exist. But the 1995 report announced with conviction that there was now a 'discernible human influence' on climate."
>
> "...a discernible human influence' was written into the 1995 summary report after the scientists themselves had gone home. Originally, the document said scientists couldn't detect a human influence on climate for sure, and they didn't know when they would. They said explicitly, 'we don't know.' That statement was deleted and replaced with a new statement that a discernible human influence did indeed exist. It was a major change."[86]

Of course, the "major change" that Dr. Crichton accurately refers to is in the IPCC's second assessment, which I also discussed in my "Greatest Hoax" speech. Dr. Crichton's book also addresses the change in terms from "global warming" to "climate change," which didn't happen in full force until after the appearance of Al Gore's global warming Hollywood hysteria film in 2006. In the novel, two of the characters, Drake and Henley, devise a way to change the rhetoric to keep money flowing in as the issue of global warming loses credibility:

"...You can't raise a dime with it, especially in winter. Every time it snows people forget all about global warming. Or else they decide some warming might be a good thing after all. They're trudging through the snow, hoping for a little global warming. It's not like pollution, John. Pollution worked. It still works. Pollution scares...people. You tell 'em they'll get cancer, and the money rolls in. But nobody is scared of a little warming. Especially if it won't happen for a hundred years."[87]

"So what you need . . . is to structure the information so that whatever kind of weather occurs, it always confirms your message. That's the virtue of shifting the focus to abrupt climate change. It enables you to use everything that happens. There will always be floods, and freezing storms, and cyclones, and hurricanes. These events will always get headlines and airtime. And in every instance, you can claim it as an example of abrupt climate change caused by global warming. So the message gets reinforced. The urgency is increased."[88]

In the "Author's Message" at the end of the book, Dr. Crichton's words are calm amid the building hysteria: "We are also in the midst of a natural warming trend that began about 1850, as we emerged from a 400-year cold spell known as the Little Ice Age"; "Nobody knows how much of the present warming trend might be a natural phenomenon"; and, "Nobody knows how much of the present trend might be man-made." And for those who were worried about impending disaster in the coming century, Dr. Crichton says, "I suspect that people of 2100 will be much richer than we are, consume more energy, have a smaller global population, and enjoy more wilderness than we have today. I don't think we have to worry about them."

In 2005, I invited Dr. Crichton to testify at a Senate hearing on the "Role of Science in Environmental Policy Making." His appearance was ridiculed by environmentalists and many Democratic Senators who dismissed his testimony because he was a science fiction writer—of course, only a few years later, those same Senators would invite the Hollywood elite to testify on Capitol Hill as experts on the dangers of global warming. Dr. Crichton was more than just a science fiction author. He was also a scientist and a medical doctor who held degrees from Harvard College and Harvard Medical School; was a visiting lecturer in Physical Anthropology at Cambridge University; and a post-doctoral fellow at the Salk Institute. His detractors didn't understand that his testimony pulled back the curtain on the manipulation of data by climate researchers, which would not be completely understood until the Climategate scandal, four years later.

What he discovered in researching the science behind climate change was that most of the research was being conducted by the same insular group of scientists without any independent verification. His message at the Senate hearing was very clear and should be endorsed by everyone who desires that public policy be set by sound science. He stated:

> In essence, science is nothing more than a method of inquiry. The method says an assertion is valid—and will be universally accepted— only if it can be reproduced by others, and thereby independently verified. The scientific method is utterly apolitical. A truth in science is verifiable whether you are black or white, male or female, old or young. It's verifiable whether you know the experimenter, or whether you don't. It's verifiable whether you like the results of a study, or you don't.

> Thus, when adhered to, the scientific method can transcend politics. Unfortunately, the converse may also be true: when politics takes precedent over content, it is often because the primacy of independent verification has been abandoned.

> Verification may take several forms. I come from medicine, where the gold standard is the randomized double-blind study. Not every study is conducted in this way, but it is held up as the ultimate goal.

In climate research, the same small group of scientists conducts the majority of the research and peer reviews each other's work. A scientist peer reviewing a colleague one year knows that that same colleague may be reviewing their work the next year. To provide a direct contrast with the research procedures in climate science, Dr. Crichton told the story of a physician who was in the middle of an FDA study of a new drug. It was a double-blind study so the different teams conducting the research were not allowed to have any contact with the other teams as they worked, so as not to contaminate the results. When this physician innocently met another researcher from a different team while waiting at the airport, they were both required to report their encounter to the FDA. As Dr. Crichton explained, "For a person with a medical background, accustomed to this degree of rigor in research, the protocols of climate science appear considerably more relaxed." As an example of this lax peer review, he specifically evaluated the work of Dr. Mann and his hockey stick.

The American climate researcher Michael Mann and his co-workers published an estimate of global temperatures from the year 1000 to 1980. Mann's results appeared to show a spike in recent temperatures that was unprecedented in the last thousand years. His alarming report received widespread publicity and formed the centerpiece of the UN's Third Assessment Report, in 2001. The graph appeared on the first page of the IPCC Executive Summary.

Mann's work was initially criticized because his graph didn't show the well-known Medieval Warm Period, when temperatures were warmer than they are today, or the Little Ice Age, when they were colder than today. But real fireworks began when two Canadian researchers, McIntyre and McKitrick, attempted to replicate Mann's study. They found grave errors in the work, which they detailed in 2003: calculation errors, data used twice, and a computer program that generated a hockeystick out of any data fed to it—even random data.

Mann's work has been dismissed as "phony" and "rubbish" by climate scientists around the world who subscribe to global warming. Some have asked why the UN accepted Mann's report so uncritically. It is unsettling to learn Mann himself was in charge of the section of the report that included his work. This episode of climate science is far from the standards of independent verification.

The hockeystick controversy drags on. But I would direct the Committee's attention to three aspects of this story. First, six years passed between Mann's publication and the first detailed accounts of errors in his work. This is simply too long for policymakers to wait for validated results. Particularly if it is going to be shown around the world in the meantime.

Second, the flaws in Mann's work were not caught by climate scientists, but rather by outsiders—in this case, an economist and a mathematician. McIntyre and McKitrick had to go to great lengths to obtain the data from Mann's team, which obstructed them at every turn. When the Canadians sought help from the NSF, they were told that Mann was under no obligation to provide his data to other researchers for independent verification.

Third, this kind of stonewalling is not unique or uncommon. The Canadians are now attempting to replicate other climate studies and are getting the same runaround from other researchers. One leading light in the field told them: "Why should I make the data available to you, when your aim is to try and find something wrong with it?"

Even further, some scientists complain the task of archiving is so time-consuming as to prevent them from getting any work done. But this is nonsense.

Dr. Crichton was right. It is nonsense that climate science is not properly peer-reviewed, it is nonsense that scientists do not share data, it is nonsense for policymakers to set policy on research which cannot be replicated, and it is nonsense to wreck economies and jobs based on this so-called research.

After the hearing, one of my staffers asked Dr. Crichton if the *State of Fear* would ever be made into a movie. He said, "No, Hollywood would never touch a film like this." He said he had lost a lot of Hollywood friends because of this book, but he felt that it still had to be written. Of course, at the time, Hollywood had just premiered *The Day After Tomorrow*, a disaster film that depicts the world's untimely demise due to our failure to "act" on global warming.

Even toward the end of his life, Dr. Crichton endured a good deal of wrath from the environmental community, and many regarded him as a traitor for portraying the errors in global warming science and exposing the shallowness of the Hollywood elite. May he rest in peace now that he too is vindicated. I invite you to read a few excerpts from *State of Fear* in Appendix B.

THE PUSH FOR CAP AND TRADE BEGINS

In 2003, when I was fighting against the hoax of the science, I was also fighting the hoax of the so-called solution to global warming: cap and trade legislation.

Cap and trade achieved essentially the same outcome as the Kyoto Treaty: it had the same $300–400 billion annual pricetag; it created the same increased energy costs; it destroyed the same number of jobs; and brought the same amount of economic pain to our country. It only bound the United States to emissions reductions so, of course, jobs would be shipped overseas to China, India, and across the border to Mexico, where they don't have cap and trade, which means that worldwide emissions would actually increase. Honestly, if I had a dollar for every time I said that Kyoto or cap and trade would have been all pain for no environmental gain, I think I would have a million dollars just from that by now.

So what they couldn't achieve through the Kyoto Treaty, they tried to achieve through legislation.

Here's how cap and trade for carbon dioxide emissions works: As the government imposes caps on emissions, it essentially establishes an artificial price for carbon. Each regulated entity may only emit a certain amount of carbon, and if it exceeds that limit, it can buy credits from other entities that are not exceeding their limits. Of course, higher emitting entities such as coal-fired power plants, would have to purchase a large number of credits to continue business as usual, and as President Obama said himself, "electricity prices would necessarily skyrocket" because these costs "would be passed on to consumers" in the form of an energy tax. Ultimately, the real losers in this scenario would have been the American people, who would have had to shoulder the largest tax increase in American history.

Of course, the philosophy behind cap and trade is that if we restrict enough supply of fossil fuels, the price will increase, and we can then simply shift to less costly alternatives. Yet this is wishful thinking. Alternatives are fine, but in most cases, they aren't widely available or commercially viable yet—certainly not in a form that can efficiently, affordably, and reliably meet our existing energy needs. How are we supposed to run this machine called America without proven and reliable sources such as oil, coal, and natural gas? The answer is we can't.

All I knew was that there was no way cap and trade was going to pass out of the Environment and Public Works Committee with me as chairman.

The first attempt to impose cap and trade came from Senators McCain and Lieberman's bill in 2003. As promised, I blocked it from moving through the Committee, so in order to have their bill brought up for a vote, they had to bypass the committee process and have it brought straight to the Senate floor. At the time, Senator McCain was living up to his reputation for being a maverick and bucking his own party. He had the support of the liberal media and the Democrats in Congress. I handled the opposition on the floor, as few Republican Senators dared to speak out against the bill. To put it simply, many of my Republican colleagues were afraid that they too would see their faces on an environmental WANTED poster.

In one lively floor debate, on October 30, 2003, Senator McCain, after quoting Ernest Hemingway, made his case that global warming was a serious threat and because of it "the snows of Kilimanjaro may soon exist only in literature." He went on to say, "These are facts. These are facts that cannot be refuted by any scientist or any union or any special interest that is weighing in more heavily on this issue than any issue since we got into campaign finance reform."

When it was my turn to speak, I said that while I appreciated the comments made by my good friend from Arizona, what he was saying was simply false. I then quoted an article from the front page of that morning's *USA Today*. It was about James Morison, a scientist with the University of Washington, who said the temperature increases and the shifts in winds and ocean currents that occurred early in the 1990s have since "relaxed" and such changes in the Arctic Circle "are not related to (global) climate change."[89] So it turned out that Senator McCain's incontrovertible facts about man-made catastrophic warming were refuted that very day.

Senator McCain immediately nabbed one of his staffers to ask why he didn't know about that article, and within moments, that staffer was rushing out the door to go find it for his boss. In the end, even though few other senators would come out publicly to oppose Senators McCain and Lieberman's bill, it was soundly defeated on the Senate floor by a vote of 55–43, with only forty-three senators supporting the measure.

Senators McCain and Lieberman, however, were determined to try again. In 2005, they reintroduced their bill, again having to go around me because they knew it had no shot at getting passed out of committee. When the 2005 Energy Policy Act (EPACT) was brought up for consideration, the senators offered their bill as an amendment. McCain arranged an agreement with the Senate Majority Leader that they would vote first on EPACT, then separately on McCain and Lieberman's amendment as a stand-alone bill, thus avoiding the committee process. It was another lonely battle on the Senate floor, as I was one of the only senators willing to openly stand in opposition. But while my colleagues may not have expressed their dislike of the bill in words, they certainly did so with their votes. McCain-Lieberman was dealt a crushing defeat on June 22 by a 60–38 vote, with only thirty-eight senators supporting the measure.

CLEAR SKIES

"The Clear Skies bill is the most aggressive presidential initiative in history to reduce power plant pollution and provide cleaner air across the country. The bill reduces emissions of sulfur dioxide, nitrogen oxides, and—for the first time—mercury from power plants by 70 percent by 2018 through expanding the successful Acid Rain Trading Program. This program, combined with the historic diesel rules being implemented by the Bush Administration, provide a national clean air strategy that will bring nearly all of the nation's counties that are not meeting clean air standards into attainment, makes the future for clean coal possible, and keeps energy affordable, reliable and secure. It is my hope that if the environmental community and my friends on the other side of the aisle are truly serious about protecting the environment they will join me and Senator Voinovich in supporting this important legislation."

—*Senate Environment and Public Works Committee Chairman James M. Inhofe, January 25, 2005*[90]

Most people are surprised to discover that while I was working to defeat cap and trade, I was also simultaneously working to pass the Clear Skies Act, which would have been the most aggressive legislation in history to improve air quality by reducing power plant emissions by 70 percent.

EPA had declared that 474 U.S. counties were in non-attainment for the new, more stringent National Ambient Air Quality Standards (NAAQS) for ozone, and that 225 counties did not meet new standards for particulate matter. Non-attainment is determined by air monitoring devices placed around cities all over the country. If any particular area violates the NAAQS standards, then they are considered to be in non-attainment, which triggers a number of regulatory requirements for that area until they can show that they have attained the standards. Such designations place a significant burden on state and municipal governments, forcing them to develop plans to reduce emissions and reach attainment by a certain date. Typically cities in nonattainment must reduce their emissions from motor vehicles (including cars and buses) or by limiting the emissions from power plants and manufacturing facilities. Failure to reach attainment can mean the rejection of any permits for new businesses in the community and the threat of losing federal highway dollars to improve the roads. Clear Skies, coupled with diesel regulations that had already been finalized by EPA, would have brought most of those counties into attainment without any additional, local controls because it would have placed most of the burden on the electricity sector, instead of state and municipal governments.

President Bush announced the Clear Skies Initiative in February 2002, and I first introduced the corresponding legislation, along with Senator Voinovich in November 2003, not long after McCain-Lieberman cap and trade legislation was defeated the first time. Just before I introduced Clear Skies, I met with President Bush to discuss the path forward. One thing that stands out in my memory from that meeting is that he told me how much he envied me for having so many kids and grandkids to go fishing with. He said that with two daughters who didn't like to fish, he'd have to wait until he had grandkids. Here was a president who could truly relate as a fellow sportsman, avid outdoorsman, and, most importantly, a family man.

After the Senate rejected cap and trade for greenhouse gases the first time, I insisted that it was time to pass legislation that would actually provide real public health and environmental benefits to the American people while preserving our economy and standard of living. Yes, Clear Skies was a cap and trade bill, but with one, crucial difference from the McCain-Lieberman bill: it focused solely on reducing three *real* air pollutants: sulfur dioxide, nitrogen oxide, and mercury (for the first time in history); it did not regulate greenhouse gases.

Global warming advocates often accused Republicans of hypocrisy for supporting cap and trade for real pollutants but not greenhouse gases.

But they confused the fact that trading mechanisms for real pollutants have achieved significant environmental benefits without harming jobs and the economy, claiming the same would happen for carbon despite every credible economic analysis showing exactly the opposite. Placing a price on carbon could wreak havoc on the economy. The reason trading programs work for real pollutants is that they are mostly emitted by large, stationary, power-generating sources, and controlled by existing technologies. Greenhouse gases, on the other hand, are emitted everywhere and by every sector of the economy, not just power plants, and they are not harmful to human health.

Because the foundation for Clear Skies was the successful Acid Rain trading program advanced by President George H.W. Bush in 1990, we knew the technology to make these reductions was viable and affordable, and that it would not raise electricity costs for consumers. On the other hand, there was no viable technology available for utilities to reduce carbon emissions. The timetables of Clear Skies achieved significant reductions that were reasonable, and the electric generating industry knew, from experience, that they could comply. Also, unlike most of our nation's environmental laws and regulations under the Clean Air Act, which have resulted in endless litigation, the Acid Rain program resulted in virtually no litigation and has achieved goals of substantial reductions in acid rain *at less than the projected cost.* Clear Skies similarly would have avoided that constant litigation. It would have improved our air by reducing utility emissions faster, cheaper, and more efficiently than the Clean Air Act as it still stands today.

The legislation went nowhere during President Bush's first term. Why would Democrats want to hand a president they hated a key legislative victory—on an issue they claim to own—before the 2004 elections?

In March 2005, Senator Voinovich and I brought the bill up again, but again Democrats blocked its consideration because it still did not regulate greenhouse gases, despite the fact that it would have made great strides in improving air quality.

The way most Democrats tell it, the Grim Reaper waits outside your door each and every day to blow toxins in your face. They would never admit that at the time I re-introduced Clear Skies in the opening days of the 109th Congress, emissions of the six main air pollutants had actually *decreased 51 percent* since 1970. That is a significant improvement over a thirty-five-year period during which U.S. gross domestic product increased by 176 percent, vehicle miles traveled increased 155 percent, energy consumption rose 45 percent, and population expanded by 39 percent. We could have built on thirty-five years of success, but politics prevailed.

Well-financed environmental NGOs such as the National Resources Defense Council and Sierra Club, or "Big Green" as I call them collectively, launched an aggressive, politically coordinated campaign against Clear Skies. The basis of their opposition was the false and utterly absurd claim that Clear Skies was a "rollback" of existing Clean Air Act provisions. This was simply not true. The real problem for Big Green was that the bill did not address greenhouse gases. And my Democrat colleagues did not like that it had been advanced by George W. Bush and me. Opponents amusingly called Clear Skies "Orwellian." The bill proposed the first-ever cap to reduce mercury emissions from the power plants by 70 percent, yet, true to form, these Big Green groups said it would allow more mercury to go into the air. Go figure. Clear Skies opponents knew that it would be more difficult to pass greenhouse gas regulations *in addition to* a three-pollutant bill. Because of this, they held it hostage, making it very clear that politics, not the environment, was the priority. Real and meaningful results for air quality were shamefully sacrificed for worthless rhetoric.

In March 2005, the bill unfortunately failed by a 9–9 tie vote in committee, thanks chiefly to Senator Lincoln Chafee, who went on to lose his Republican-held seat in the Senate despite his calculated opposition to appease the Big Green machine. Senator Voinovich and I made a number of compromises to advance the legislation, and even agreed to postpone scheduled markups three times in a good faith effort to work with opponents to reach a consensus. In the end, after several months of markup delays, twenty-four hearings, five years, and more than 10,000 pages of modeling data, the Democrats succeeded in killing Clear Skies. If anyone was denied an environmental victory, it was the American people.

3

THE BUILD-UP OF ALARMISM

CELEBRITIES AND THE "CLIMATE CRISIS"

Most people know that I'm not much of a movie-goer, and I can't tell one Hollywood star from another. Let's just say that if you would have told me a few years earlier that I would be giving PowerPoint presentations that included quotes from the likes of Leonardo DiCaprio, Barbra Streisand, and John Travolta, I wouldn't have believed you. But the alarmists had taken the fight to pop culture, so I was determined to join in the fray.

After suffering two overwhelming defeats in Congress with the

McCain-Lieberman cap and trade bill, the environmental left realized that they had to go bigger than Capitol Hill and take their alarmist message of fear directly to the people. Conveniently, they had Hollywood and the mainstream media on their side. It was a marriage made in heaven—or so they thought at the time.

By 2006, the American public was inundated with an unprecedented parade of environmental alarmism though films and celebrities, all portending a future of natural disasters, wars, displacements, and plagues. They were also going full force with the major news organizations, which had completely dismissed any pretense of balance and objectivity on climate coverage and instead focused squarely on promoting global warming advocacy. Basically, everywhere you turned you were bombarded with the message that the end of the world was nigh.

There I was, this not particularly glamorous Senator from Oklahoma, standing up against these young, beautiful celebrities who had come together to proclaim with one voice that the science was settled and we had to act immediately to avoid climate catastrophe.

But they had a little bit of a problem: their "solutions" were absurd. Cameron Diaz stood on a stage at one particular climate concert and proclaimed that women could mitigate global warming by turning off the water in the shower when they shaved their legs;[91] Laurie David, an activist with strong ties to Hollywood, said that we needed to change our standard light bulbs to fluorescents to avert crisis;[92] and of course the most famous "solution" came from Sheryl Crow who said a "limitation should be put on how many squares of toilet paper can be used in one sitting."[93] That is, unless there was a serious emergency, in which case more than one square would be acceptable. She later claimed that she was just joking, but as I said on Fox News, "I'm just glad she didn't define what that serious emergency was."

There is certainly nothing wrong with the desire to conserve energy, but if we were indeed facing a crisis of the proportions they predicted, the suggestion that we can save the world one square of toilet paper at a time was just absurd. In the midst of all this hysteria, my message remained calm, consistent and clear: the science is not settled. Their so-called "solutions" were only symbolic and would have no actual impact on the climate.

Looking back on these years, I'm sure that many people have probably forgotten just how prevalent global warming was in the mainstream;

it seemed as if a day didn't go by without hearing about the impending doom. Now it is all but forgotten, but they will be back.

HOLLYWOOD HYSTERIA

Of course, of all the efforts to push alarmism, nothing was bigger than Al Gore's documentary, *An Inconvenient Truth*, which rocketed the global warming movement into pop culture. Suddenly the alarmists had the hero they were craving: Gore was now the "climate prophet" or the "Goricle" as journalist Howard Fineman once called him.[94] Katie Couric famously said that Gore was a "Secular Saint,"[95] and Oprah Winfrey said that he was the "Noah"[96] of our time. This is what I'm talking about when I say that it was a religion to them.

But if Gore was a prophet, he was certainly a prophet of doom. Even the trailer to the film flashes ominous images of destruction accompanied with phrases such as "Nothing is scarier than the truth;" "By far the most terrifying film you will ever see;" it's a "film that has shocked audiences worldwide." Given the twenty significant scientific errors I found in Gore's film, it was clear that the science was secondary to the primary goal of promoting fear and pushing the message that the debate was over and the science was settled.

Gore chose to ignore those inconvenient scientists such as Richard Lindzen of MIT who put it well when he said, "A general characteristic of Mr. Gore's approach is to assiduously ignore the fact that the earth and its climate are dynamic; they are always changing even without any external forcing. To treat all change as something to fear is bad enough; to do so in order to exploit that fear is much worse."[97]

The architect behind bringing the global warming alarmism message to Hollywood was Laurie David, an environmental activist with close ties to Hollywood. She convinced Gore to turn the PowerPoint presentation that he was giving across the world into a documentary and became a co-producer of the film. She was also behind numerous one-sided fear mongering documentaries that aired on TBS, CNN, HBO, and even Fox News. She teamed up with her husband to produce a two-hour comedy on TBS called "Earth to America," which included numerous stars—most notably Will Ferrell—in a hilarious spoof of President George Bush complaining that liberals were trying to make him "look bad" by "using such things as facts and scientific data."[98]

Later in 2007, Leonardo DiCaprio followed suit and completely tossed objective scientific truth out the window in a documentary scarefest called *The 11th Hour*. Like Gore and David, DiCaprio refused to acknowledge any scientists who disagreed with his dire vision of the future of the Earth. In fact, his film featured physicist Stephen Hawking making the unchallenged assertion that "the worst-case scenario is that Earth would become like its sister planet, Venus, with a temperature of 250 [degrees] centigrade."[99] In other words, worst-case scenarios pass for science in Hollywood—in fact, they are preferred. DiCaprio was not shy about stating that this was the purpose of the film. As he said, "I want the public to be very scared by what they see. I want them to see a very bleak future."[100] They may have been too scared even to attend as the film itself was apparently a miserable flop.

While the target was essentially everyone, the alarmists were particularly focused on planting the seed of fear in the young. DiCaprio announced his goal was to recruit young eco-activists to the cause: "We need to get kids young," He said in an interview with *USA Weekend*.[101] Laurie David also coauthored a children's global warming book with Cambria Gordon for Scholastic Books titled, *The Down-To-Earth Guide to Global Warming*. David made it clear the purpose of her book was to influence young minds when she wrote in an open letter to her children, "We want you to grow up to be activists."[102]

After a successful campaign to have *An Inconvenient Truth* shown in classrooms around the nation, the alarmists were unfortunately having the impact they wanted on children: Nine-year-old Alyssa Luz-Ricca was quoted in *The Washington Post* on April 16, 2007, as saying: "I worry about [global warming] because I don't want to die."[103]

I remember one morning, when I was grocery shopping back home in Tulsa with Kay, a young mother introduced herself and told me how worried she was that her child had been made to watch *An Inconvenient Truth* in several classes at school without being told there was another side to the story. These poor kids were being bombarded with a scientifically unfounded doomsday message designed to create fear, nervousness, and ultimately recruit them to liberal activism.

MEDIA MANIA: COOLING OR WARMING CATASTROPHE?

Through it all, the mainstream media had nothing but praise for these efforts. I almost don't blame the major news outlets for jumping on global warming hysteria with a vengeance. Let's be honest: catastrophe sells news. So rather than focus on the hard science of global warming, and looking at all the possibilities, the media instead became prime advocates for hyping scientifically unfounded climate alarmism.

Such tactics had certainly worked to their advantage before. Take, for example, a quote from the *New York Times* reporting fears of an approaching ice age: "Geologists Think the World May be Frozen Up Again."[104] That sentence appeared more than one hundred years ago in the February 24, 1895, edition of the *New York Times*. Then a front-page article in the October 7, 1912, *New York Times*, just a few months after the *Titanic* struck an iceberg and sank, declared that a prominent professor "Warns Us of an Encroaching Ice Age."[105] The very same day in 1912, the *Los Angeles Times* ran an article warning that the "Human race will have to fight for its existence against cold."[106] An August 10, 1923, a *Washington Post* article declared: "Ice Age Coming Here."[107]

By the 1930s, the media took a break from reporting on the coming ice age and instead switched gears to promoting global warming: "America in Longest Warm Spell Since 1776; Temperature Line Records a 25-year Rise," stated an article in the *New York Times* on March 27, 1933.[108] The media of yesteryear was also not above injecting large amounts of fear and alarmism into their climate articles. An August 9, 1923, front-page article in the *Chicago Tribune* declared: "Scientist Says Arctic Ice Will Wipe Out Canada."[109] The article quoted a Yale University professor who predicted that large parts of Europe and Asia would be "wiped out" and Switzerland would be "entirely obliterated." A December 29, 1974, *New York Times* article on global cooling reported that climatologists believed "the facts of the present climate change are such that the most optimistic experts would assign near certainty to major crop failure in a decade."[110] The article also warned that unless government officials reacted to the coming catastrophe, "mass deaths by starvation and probably in anarchy and violence" would result. In 1975, the *New York Times* reported that

"A major cooling [was] widely considered to be inevitable."[111] These past predictions of doom have a familiar ring, don't they?

During the latest global warming craze, the image that stands out the most is the *Time* magazine cover picturing the last remaining polar bear standing on the one remaining ice cube under the heading "Be Worried,

Be Very Worried."[112] I joked that Americans *should* be very worried—about such shoddy journalism. After more than a century of alternating between global cooling and warming, one would think that this media history would serve a cautionary tale for today's voices in the media and scientific community who were promoting yet another round of eco-doom.

After I presented the media's one-hundred-year-history of embarrassing climate change reporting in a speech on the Senate floor in October 2006, *Newsweek* magazine Senior Editor Jerry Adler issued a one-thousand-word correction[113] for its 1975 story on the dangers of global cooling that, as the article said, "may portent a drastic decline in food production—with serious political implications for just about every nation on Earth."[114] It took them thirty-one years to admit that that these statements were "so spectacularly wrong about the near-term future."[115] But Adler still wasn't willing to put the full blame on *Newsweek* for this incredible gaffe. As he said, "The story wasn't 'wrong' in the journalistic sense of 'inaccurate.'"[116] His justification was that "Some scientists indeed thought the Earth might be cooling in the 1970s, and some laymen—even one as sophisticated and well-educated as Isaac Asimov—saw potentially dire implications for climate and food production."[117] Yet, he was still unwilling to admit that what the media now says about global warming could be wrong, as it was in the 1970s. So I'm guessing that if it takes thirty one years for *Newsweek* to admit its mistakes, we can expect them to recant their latest global warming scare around October 2037.

But Adler was right in one respect. He also said, "All in all, it's probably just as well that society elected not to follow one of the possible solutions mentioned in the *Newsweek* article: to pour soot over the Arctic ice cap, to help it melt."[118]

Newsweek was not the only publication that responded to me calling them out over their past climate debacles. The *New York Times* also weighed in with an October editorial:

> We do not expect Mr. Inhofe to see the light—or feel the heat—any time soon. He and his staff are serious collectors of opposition research. But the essence of his strategy is to seize upon a mistaken or over-blown story to try to undermine the broad consensus. If that fails, he can always question his opponents' politics and motives, as with his insinuations that environmentalists dreamed the whole thing up to scare people and raise money.[119]

In other words, even though the *New York Times* was completely wrong about global cooling in the 1970s, they were outraged that I would have the audacity to question the validity of their claim that there is a "broad consensus" among scientists regarding global warming today.

Even my favorite global warming reporter, Miles O'Brien, wasn't too happy when I challenged him on his past climate cooling gaffes. Miles was great: so many extremists were mad all the time, but Miles always smiled, even when he was cutting my guts out, and I appreciated that. Here's the exchange in one of our typically lively television interviews:

INHOFE: And I wonder also, Miles, it wasn't long ago—you've got to keep everyone hysterical all the time. You were the one that said another ice age is coming just twelve years ago.

O'BRIEN: I said that? I didn't say that.

INHOFE: You didn't say that. Let me quote you...

O'BRIEN: No, no, no. I'd be willing to tell you there are stories like that. But there's not...

INHOFE: ...quote you so I'll be accurate. I don't want to be inaccurate.

O'BRIEN: All right, go ahead.

INHOFE: You said, in talking about a shift that was coming—you said, "If the Gulf Stream were to shift again, the British Isles could be engulfed in polar ice and Europe's climate could become frigid. [From CNN Transcript titled Scientists Research the Rapidity of the Ice Age dated December 19, 1992.]" That's another scary story.

O'BRIEN: But that also is a potential outgrowth of global warming when you talk about the ocean currents being arrested. This is "The Day After Tomorrow" scenario that we're talking about.

So global cooling was actually global warming. They were determined to have their catastrophe no matter what.[120]

MEDIA MANIA OVER GORE

Not surprisingly, Al Gore had the full backing of the media to promote his movie, and leading the cheerleading charge was none other than the Associated Press. On June 27, 2006, an AP article written by Seth Borenstein boldly declared: "Scientists give two thumbs up to Gore's movie."[121] The article states that the top scientists were giving his movie five stars for accuracy and that its prospects of "a flooded New York City, an inundated Florida, more and nastier hurricanes, worsening droughts, retreating glaciers and disappearing ice sheets" were mostly accurate, with only some minor adjustments.

Of course, the AP chose to ignore the scores of scientists who have harshly criticized the science in Gore's movie. I said that the AP should release the names of the "more than 100 top climate researchers" that they attempted to contact, and the nineteen scientists who gave Gore "five stars for accuracy." Most importantly, the AP chose to ignore Gore's reliance on the "hockey stick" by Dr. Michael Mann, which claims that temperatures in the Northern Hemisphere remained relatively stable over nine hundred years, then spiked upward in the twentieth century, and that the 1990s were the warmest decade in at least one thousand years. Only a week before the AP article was published, the National Academy of Sciences report dispelled Mann's oft-cited claims by reaffirming the existence of both the Medieval Warm Period and the Little Ice Age.[122] That's when I declared on the Senate floor that the hockey stick was broken. Yet, this highly significant breaking news is not even acknowledged in this article which instead features only glowing praise from the likes of Robert Corell, chairman of the worldwide Arctic Climate Impact Assessment group of scientists, who said he was "amazed at how thorough and accurate" it was; he was "blown away" and "could find no error."[123]

WHAT BALANCE?

This AP article epitomizes the attitude of many journalists around that time: balance in global warming reporting simply wasn't valued, and many were not at all afraid to admit that openly. The quote that perhaps encapsulates the global warming mania in the mainstream media most comes from Bill Blakemore with ABC News, who explained on August 30, 2006, "After extensive searches, ABC News has found no such [scientific] debate" on global warming.[124]

ABC News put forth its best effort to secure its standing as an advocate for climate alarmism when the network put out a call for people to submit their anecdotal global warming horror stories in June 2006 for use in a future news segment.[125] Then, in July of that year, the Discovery Channel presented a documentary on global warming, narrated by former NBC anchor Tom Brokaw. You don't have to take my word on the program's overwhelming bias; a Bloomberg News TV review noted, "You'll find more dissent at a North Korean political rally than in this program" because of its lack of scientific objectivity.[126]

Brokaw, who had affiliations with the Sierra Club, lavishly praised Gore's film as "stylish and compelling," called the science behind catastrophic human caused global warming "irrefutable" and specifically presented climate alarmist James Hansen's views as unbiased. Of course, he failed to note Hansen's quarter-million-dollar grant from the partisan Heinz Foundation or his endorsement of Democrat Presidential nominee John Kerry in 2004 and his role promoting former Vice President Gore's Hollywood movie.

Brokaw, however, did find time to impugn the motives of scientists skeptical of climate alarmism when he featured paid environmental partisan Michael Oppenheimer of the Environmental Defense Fund, accusing skeptics of being bought out by the fossil fuel interests. Whenever the media asked me how much I have received in campaign contributions from the fossil fuel industry, my unapologetic answer was "not enough"— especially when you consider the millions partisan environmental groups pour into political campaigns.

Former Colorado state climatologist and professor emeritus of atmospheric sciences at Colorado State University, Senior Research Scientist in the Cooperative Institute for Research in Environmental Sciences (CIRES), and Senior Research Associate in the Department of Atmospheric and Oceanic Sciences (ATOC) Roger Pielke Sr., viewed an advance copy of Brokaw's special and declared that it contained "errors and misconceptions" and "relied on just a few scientists with a particular personal viewpoint on this subject which misleads the public on the broader view that is actually held by most climate scientists."[127]

In March of that year, *60 Minutes* profiled NASA scientist and alarmist James Hansen, who was once making allegations of being censored by the Bush administration.[128] Many at that time pointed out the irony of a man who claimed to be censored, yet was appearing frequently on every major network.

In this segment, objectivity and balance were again tossed aside in favor of a one-sided glowing profile of their hero Hansen. It made no mention of Hansen's ties to quarter-of-a-million-dollar grant from the left-wing Heinz Foundation run by Teresa Heinz Kerry. Neither did *60 Minutes* inform viewers that Hansen appeared to concede in a 2003 issue of *Natural Science* that the use of "extreme scenarios" to dramatize climate change "may have been appropriate at one time" to drive the public's attention to the issue.[129]

To put the severity of this lack of balance in perspective, one of Laurie David's one-sided documentaries, featuring David herself as well as Robert Kennedy Jr., was even aired on Fox News in November 2005.[130] As the Fox News promotional article said, they were "committed to teaching everyday Americans and the rest of the world about what can be done to cut down on greenhouse gases that threaten our children's future." The surprise of seeing such a one-sided documentary on Fox was not lost on the rest of the mainstream media. The *Los Angeles Times* reported, "When Dan Becker, director of the Sierra Club's global-warming program, got a call from the network this summer asking him to be interviewed for the documentary, he initially thought it was a prank call. 'I asked whether they were joking,' said Becker, who participated and noted that he was ultimately impressed by the network's questions."[131] When I found out that Fox News was moving forward with this unbalanced documentary, I immediately called Roger Ailes, who is the president of the network, and he conceded that it was not a journalistically balanced approach. He agreed then to feature another documentary that tells the other side of the story.

Perhaps the funniest example of the lack of balance in environmental reporting comes from a particular ad by CBS News seeking a "Hip Environmental Reporter" with no need to have "knowledge of the enviro beat" but must be "funny, irreverent and hip, oozing enthusiasm and creative energy." This employee must be "vibrant" and bring "a dash of humor to our coverage." I actually gave them credit for coming out and admitting that being "hip" was the most important thing for selling what was then the "cool" environmental agenda.[132]

LAST POLAR BEAR STANDING ON
THE LAST CUBE OF ICE

Through it all, the last polar bear standing on the last cube of ice remained the image seared in everyone's mind from the media of why we should be worried—very worried about global warming.

On August 3, 2006, the *New York Times* ran an op-ed by Bob Herbert called "Hot Enough Yet?" which claimed that polar bears were "drowning because they can't swim far enough to make it from one ice floe to another."[133] Then a Reuters article dated September 15, 2006, by correspondent Alister Doyle, quoted a visitor to the Arctic who claimed that he saw two distressed polar bears: "one of [the polar bears] looked to be dead and the other one looked to be exhausted."[134] Importantly the article did not state that the bears were actually dead or exhausted, rather that they "looked" that way. At the time, I asked, have we really arrived at the point where major news outlets in the United States are reduced to analyzing whether or not polar bears in the Arctic *appear* to be tired? How does reporting like this get approved for publication by the editors at *Reuters*? Where was the rigorous scientific discussion?

What was missing from this Reuters news article was the fact that according to biologists who actually study the animals for a living, polar bears were doing quite well. Biologist Dr. Mitchell Taylor from the Arctic government of Nunavut, a territory of Canada, refuted these claims in May of that year when he noted: "Of the 13 populations of polar bears in Canada, 11 are stable or increasing in number. They are not going extinct, or even appear to be affected at present."[135] At that stage, it was clear that it was more interesting to hype up a polar bear catastrophe in the media than to engage in the basic tenets of balance in journalism.

In September 2006, a report based on unproven computer models found that polar bear populations were allegedly going to be devastated by 2050 due to global warming.[136] The report was issued as part of the U.S.

Fish and Wildlife Service's consideration of listing the polar bear under the Endangered Species Act. It was a classic case of reality versus speculation, when you consider that the Fish and Wildlife Service estimated that the polar bear population was at 20,000–25,000 bears that year, whereas in the 1950s and 1960s, estimates were as low as 5,000–10,000 bears. That same month, Dr. Taylor, referring to the Fish and Wildlife computer modeling report rightly said, "It is just silly to predict the demise of polar bears in 25 years based on media-assisted hysteria."[137]

The report with unproven computer models from the Fish and Wildlife Service, of course, was one of the key factors that led to the decision to list the polar bear as threatened under the Endangered Species Act later in May 2008, despite the fact that actual data was showing that the number of polar bears was not in decline.[138]

THE DREADED DENIERS

Why would these new outlets blatantly ignore the basic tenets of journalism, objectivity, and balance? A reporter at the time for CBS, Scott Pelley, provided a telling answer. According to him, excluding scientists skeptical of global warming alarmism was justified because skeptics are the equivalent of "Holocaust deniers."[139] Similarly, Robert F. Kennedy Jr. said at one of Gore's LIVE Earth concerts, "This is treason. And we need to start treating them as traitors."[140] Dave Roberts of *Grist* openly called for Nuremberg-style trials for climate skeptics.[141]

There is one particular story, however, that stands out. Heidi Cullen who, at the time, hosted a weekly global warming program called, *The Climate Code,* wrote on her blog on the Weather Channel Web site that meteorologists should be stripped of their scientific certification if they express skepticism about predictions of man-made catastrophic global warming. As she said:

> If a meteorologist can't speak to the fundamental science of climate change, then maybe the AMS shouldn't give them a Seal of Approval. Clearly, the AMS doesn't agree that global warming can be blamed on cyclical weather patterns.... It's like allowing a meteorologist to go on-air and say that hurricanes rotate clockwise and tsunamis are caused by the weather. It's not a political statement...it's just an incorrect statement.[142]

With so much wrath directed at skeptics, you can imagine the names they called me. Let's just say that if Al Gore was a secular saint, a Noah figure, and a prophet, I was dubbed at worst as the devil and at best as the "Doubting Thomas," or as the *New York Times* put it, the "Doubting Inhofe."[143] In a response to that particular *New York Times* editorial, Debra Saunders of the *San Francisco Chronicle* wrote a very clever column called "Inhofe the Apostate," which made much of the global warming "religious" rhetoric:

> The *Times'* focus was on Inhofe's refusal to bow to "the consensus among mainstream scientists and the governments of nearly every industrialized nation concerning man-made climate change." That is, Inhofe has had the effrontery to challenge elite orthodoxy. Or, as the editorial put it, Inhofe "has really buttressed himself with the will to disbelieve."
>
> Get thee away, Satan.
>
> "I see a sense of desperation that I haven't seen before," Inhofe told me by phone Thursday, "and frankly I'm enjoying it."
>
> CNN's Miles O'Brien also challenged Inhofe in a similar vein. O'Brien cited the NAS study, then assailed Inhofe with quotes from notable Republicans—President Bush, Gov. Arnold Schwarzenegger and Rep. Chris Shays of Connecticut—who recognize global warming. Note that Schwarzenegger gets into global-warming heaven just for believing, despite his four Hummers and use of a private jet.[144]

Hollywood and the mainstream media called me every name in the book: the guy who "thinks global warming is debunked every time he drinks a slushy and gets a brain freeze,"[145] "the noisiest climate skeptic,"[146] "banged on the head too many times,"[147] the "Senate's resident denier bunny,"[148] "Traitor,"[149] "Dumb," "crazy man," "science abuser,"[150] "Holocaust denier," "villain of the month," "hate-filled," "warmonger," "Neanderthal," "Genghis Khan," and "Attila the Hun." Later in 2010, I had the distinction of being included in *Rolling Stone's* list of the "Planet's Worst Enemies." My only disappointment was that I should have been No. 1, not No. 7.[151]

During those years, I couldn't help but wonder: if the advocates for global warming alarmism were so confident they had the broad consensus and the science was settled and the evidence was overwhelming, why were they so determined to silence me?

A MEMORABLE INTERVIEW: MILES O'BRIEN AND THE ONE-MAN TRUTH SQUAD

The interview with CNN's Miles O'Brien that Debra Saunders mentioned in her column was a particularly memorable exchange that is one of the best examples of the alarmists' narrative that everyone was against me and I was the only one not buying everything they were telling me.

On September 25, 2006, I gave a speech on the Senate floor called "Hot & Cold Media Spin Cycle: A Challenge to Journalists Who Cover Global Warming" which exposed the glaring media bias on global warming. A few days later, on September 28, 2006, CNN ran a segment criticizing my speech and attempting to refute the scientific evidence I presented to counter media-hyped climate fears.[152] CNN reporter Miles O'Brien inaccurately claimed that I was "too busy" to appear on his program. They were told simply that I was not available on Tuesday or Wednesday and that I preferred to do the segment live.

On January 31, 2007, Miles and I finally had the chance to argue live on national TV. Right away Miles began building up the idea that I was completely alone amid the overwhelming "consensus." He began with the science:

> **O'BRIEN:** Let's talk about the science first. We've got a big report coming out, this United Nations report, 2,500 of the world's leading scientists. It's being called a smoking gun report with a link between humans and global warming. Let's listen to what one of the leading scientists has to say about it.
>
> **JAMES HANSEN, DIR., GODDARD INST. FOR SPACE STUDIES:** The human link is crystal clear. There is no question, the increase from 280 to 380 parts per million in CO_2 is due to the burning of fossil fuels.
>
> **O'BRIEN:** That's James Hansen, one of the leading climate scientists. He says it's crystal clear. What do you say?[153]

I said that's James Hansen, who is paid $250,000 by the Heinz Foundation. I told Miles he was wrong to say that the report was going to come out that Friday. He would not be reading the actual report but the *Summary for Policymakers*—the report would not surface until weeks later. Of course, the *Summary* is all the media and members of Congress were ever going to look at. They were never going to talk about anything else—and this *Summary* was written by politicians, *not scientists*. In fact, the IPCC's own guidelines explicitly state that the scientific reports have to be "change[d]" to "ensure consistency with" the politically motivated *Summary for Policymakers*: "Changes (other than grammatical or minor editorial changes) made after acceptance by the Working Group or the Panel shall be those necessary to ensure consistency with the *Summary for Policymakers* or the Overview Chapter."[154]

I wasn't alone in expressing my concerns with the UN mandating that scientific work be altered to fit its political agenda. As Steve McIntyre, who had debunked the hockey stick graph, said:

> So the purpose of the three-month delay between the publication of the (IPCC) *Summary for Policymakers* and the release of the actual WG1 (Working Group 1) is to enable them to make any 'necessary' adjustments to the technical report to match the policy summary. Unbelievable. Can you imagine what securities commissions would say if business promoters issued a big promotion and then the promoters made the 'necessary' adjustments to the qualifying reports and financial statements so that they matched the promotion.[155]

Harvard University Physicist Lubos Motl also slammed the UN saying, "These people are openly declaring that they are going to commit scientific misconduct that will be paid for by the United Nations. If they find an error in the summary, they won't fix it. Instead, they will 'adjust' the technical report so that it looks consistent."[156]

Another aspect Miles made a point of challenging me on was my religion. As Miles correctly noted, I take my religion seriously—I always say I'm a Jesus guy—so why wasn't I buying into what evangelicals such as Rev. Richard Cizik were saying? In May 2006, *Vanity Fair* featured a photo of Cizik dressed like Jesus and walking on water—another savior figure in this "religion" of global warming. Cizik was being sponsored by many environmentalist groups who were trying to break into the National Association of Evangelicals. Fortunately, evangelicals had already rejected him, and for good reason. In May 2006, in a speech to the World Bank, Cizik said, "I'd like to take on the population issue… We need to confront population control and we can—we're not Roman Catholics after all—but it's too hot to handle now."[157] So those were his true views, and he did not represent evangelicals.

Miles wasn't the only one who was pushing this idea that religious leaders were buying into global warming fears. In our joint interview with Larry King on January 31, 2007, Senator Boxer said that in addition to the Big Oil companies that were calling her to express their support of cap and trade, "evangelicals are coming to me. So a consensus is building and my dear friend Jim Inhofe is just being left all alone."[158] Later on October 13, 2009, when she was announcing hearings on her own cap and trade bill, she said, "I can report that Evangelical groups and other religious communities have expressed their commitment to help us move the bill quickly."[159]

But this was misleading: directly countering the views of Cizik and those in the mainstream media who were pushing the idea that evangelicals were on board with global warming hysteria, in 2000, the Cornwall Alliance released the Cornwall Declaration which reminds us, "We should respect creation and be wise stewards, but we must be careful not to fall into the trap of secular environmentalists who believe that man is an afterthought on this earth, and is principally a polluter of it." Later in 2009, the Cornwall Alliance's evangelical arm issued a statement called "An Evangelical Declaration on Global Warming," which stated that "recent progress in climate research suggests that:

1. Observed warming and purported dangerous effects have been overstated.

2. Earth's climate is less sensitive to the addition of CO_2 than the alleged scientific consensus claims it to be, which means that climate model predictions of future warming are exaggerated.

3. Those climate changes that have occurred are consistent with natural cycles driven by internal changes in the climate system itself, eternal changes in solar activity, or both.[160]

I do not pretend to be a biblical scholar, but I have read a lot of work by scholars on the topic of man's relation to creation and what stewardship means from a biblical perspective. It seems to me that we should make use of the resources God has given us, and remember that it is God, not God's creation, that should be praised, as exemplified in Romans 1:25: "They gave up the truth about God for a lie, and they worshiped God's creation instead of God who will be praised forever. Amen."[161]

I have always greatly admired Bill Bright, the founder of Campus Crusade for Christ, who was a close friend of mine from 1984 until his death. He wrote a daily devotional book called *Promises*, which I have read nearly every day for twenty-five years. Many times during my global warming fight, I turned to Day 36 of *Promises* which features one of my favorite Bible verses, Genesis 8:22:

As long as the earth remains

there will be springtime and harvest,

cold and heat, winter and summer,

day and night.

The devotional associated with the verse in *Promises* is a wonderful story about how it is God who "maintains the seasons":

On his way to a country church on Sunday morning, a preacher was overtaken by one of his deacons.

"What a bitterly cold morning," the deacon remarked. "I am sorry the weather is so wintry."

Smiling the minister replied, "I was just thanking God for keeping His Word."

"What do you mean?" the man asked with a puzzled look on his face.

"Well," the preacher said, "more than 3,000 years ago God promised that cold and heat should not cease, so I am strengthened by this weather which emphasizes the sureness of His promises."[162]

And this is what a lot of alarmists forget: God is still up there, and He promised to maintain the seasons and that cold and heat would never cease as long as the earth remains.

Miles also asked me why I remained obstinate when corporate heavyweights were coming to Capitol Hill to testify that they could compete in a "greenhouse gas constrained world." As I said to Miles, if anyone thinks these corporations are willing to reduce their emissions out of the goodness of their hearts, think again. These companies were climate profiteers that would gain market share against their competitors while the economy flattens and jobs are sent to China. At that time, ten companies, ranging from Duke to GE, announced they were joining together to create the Climate Action Partnership, but it was abundantly clear that they would individually profit from this plan. As I said to Miles, coal is responsible for over 50 percent of our electricity in America. These companies have nuclear, hydroelectric, solar, and wind, so it was to their financial advantage to do away with coal. The biggest losers in the plan would not be big businesses, but the American people, who would have to foot the bill for these companies' profits under this cap and trade regime. If I were on the board of directors of GE, who is making solar equipment and wind turbines, I'd say, "Sure, let's jump on this bandwagon. We'd make a fortune."

At the time, Republicans had begun to feel the heat of the global warming alarmism campaign, and Miles couldn't wait to ask me why I remained skeptical when my Republican counterparts such as Senator John McCain had proclaimed, "I believe climate change is real. I believe that we need to act as quickly as possible." But it was the State of the Union address by President George W. Bush that was the defining moment for the media. When President Bush said America was addicted to oil and that global warming was a serious problem, they saw this as a victory. According to them, Bush had caved and I was just being stubborn among my Republican colleagues with my insistence that the science wasn't settled. Miles also used the Bush EPA's Web site as further evidence:

O'BRIEN: First two lines there are, "According to the National Academy of Sciences, the Earth's surface temperature has risen by about a degree Fahrenheit in the past century, with accelerated warming during the past two decades. There is new and stronger evidence that most of the warming over the last fifty years is attributable to human activities." This is the Environmental Protection Agency from the Bush administration saying that. I mean, there is a lot of consensus here, isn't there?[163]

Later in 2008, I'll never forget the vision of Newt Gingrich sitting on a couch in front of the Capitol holding hands with Nancy Pelosi, saying while he and Pelosi rarely agree, "We do agree our country must take action to address climate change," and "If enough of us demand action from our leaders, we can spark the innovation that we need." Al Gore's Alliance for Climate Protection sponsored the ad. I applaud Newt for saying on October 9, 2011, three years after the couch episode, "That is probably the dumbest single thing I've done in recent years."[164]

But I told Miles that he was wrong to say that I was alone in this particular sense: I reminded him that the last time we had a vote on cap and trade, led by John McCain, we won 60–38. So there were fifty-nine other senators who agreed with me on that point, even if they weren't willing to say it. Then I went on to explain that even if it's true that the planet is warming due to anthropogenic gases, if every developed nation signed on to the Kyoto Treaty, it would only reduce the earth's temperature by 0.06 degrees Celsius by 2050.

His reaction was very interesting: "Well, we're not talking about Kyoto…We're not talking about Kyoto. We're just talking about whether global warming is real."[165]

That last comment from Miles pretty much encapsulates the entire fear campaign from about 2005 to 2007. It was completely centered on drumming up terror: what they didn't want to acknowledge was whether it was the Kyoto Treaty or cap and trade, whatever "solution" there was to this so-called "problem" was about as effective as demanding that every American use only one square of toilet paper in one sitting. Indeed, cap and trade—the greatest tax increase in American history—was actually more terrifying than the catastrophe they were perpetrating; and for them, that was a serious problem.

A HEATED HEARING

In December 2006, just before I had to hand the gavel of the Environment and Public Works Committee over to the new Chairman, Senator Boxer, we held one of my favorite hearings on how the mainstream media was over-hyping the coverage of global warming to scare the public. One of our guests, Paleoclimate researcher Bob Carter of Australia's James Cook University, who has had over one hundred papers published in revered scientific journals, said that "there is huge uncertainly in every aspect of climate change." He explained:

> If you look at the ice core records, you will discover that yes, changes in carbon dioxide are accompanied by changes in temperature, but you will also discover that the change in temperature precedes the change in carbon dioxide by several hundred years to a thousand or so years. Reflect on that. And reflect on when you last heard somebody say that they thought lung cancer caused smoking. Because that is what you are arguing if you argue on the glacial time scale that changes in carbon dioxide cause temperature changes. It is the other way around.[166]

Carter also made the important point that uncertainty is even present in the way scientists make their claims. And that was something I always found ironic about the debate. They were always going on about how the science is settled and there was unequivocal certainty of catastrophe, but their rhetoric, as Carter said, was completely couched in words like "if, could, may, might, probably, perhaps, likely, expected, projected":

> Wonderful words. So wonderful, in fact, that environmental writers scatter them through their articles on climate change like confetti. The reason is that—in the absence of empirical evidence for damaging human-caused climate change—public attention is best captured by making assertions about 'possible' change. And, of course, using the output of computer models in support, virtually any type of climatic hazard can be asserted as a possible future change.[167]

David Deming, a geophysicist from the University of Oklahoma, also echoed these concerns about the media saying:

> Every natural disaster that occurs is now linked [by the media] with global warming, no matter how tenuous or impossible the connection. As a result, the public has become vastly misinformed on this and other environmental issues.[168]

I found the entire discussion compelling but I noticed there was one person who was less than riveted. In the back of the room, there was Miles O'Brien with his head on the press table, fast asleep.[169] Oh well, I thought, there wasn't much hope of him converting anyway.

HUNGRY FOR THE TRUTH

The American people had been fed an unprecedented diet of hysterical rhetoric: they were hungry for the truth. So amid the deafening noise of Hollywood and the media, I found two powerful forums where I could bypass the mainstream media and reach out directly to the public: talk radio and my Environment and Public Works Committee Web site—a Web site that launched one the first blogs and YouTube channels in Congress.

My appreciation for talk radio goes back many years. During the height of the depression, I had a great experience that I would not fully understand for several decades. My father worked with, and came through the depression with, a sports announcer from WHO Radio in Des Moines by the name of Ronald Reagan, or "Dutch Reagan" as he was called then. He was considered family at the time and, several years later after we moved to Tulsa, we thought of him differently as he became an important political figure. I was a small child at the time and my father worked most every day of the week. We rarely went to movies, but whenever there was a Dutch Reagan movie, we would go without fail. I recall one night driving six hours round trip to Duncan, Oklahoma, just to see a Dutch Reagan movie.

From watching Reagan throughout his career, I learned about the power of talk radio to reach a wide audience—I knew it would be one of the best ways to get my message about global warming far and wide. Whether through the Steve Malzberg show in New York City or the Steve Hennen show in North Dakota, I have appreciated all the opportunities to subvert the mainstream media. Now with Twitter and Facebook, it has been easy to alert people across the country whenever

I'm going to speak on a local Oklahoma radio show.

Now there is also a large global warming community online that offers diverse perspectives on the ongoing debate. There are several top skeptic sites that I recommend, but to name a few I would suggest R.A. Pielke Sr., Colorado State Climatologist Emeritus and Senior Research Scientist CIRES University of Colorado at Boulder (http://pielkeclimatesci.word-press.com), his son Roger Pielke Jr., professor of environmental studies at the Center for Science and Technology Policy Research at the University of Colorado at Boulder (http://rogerpielkejr.blogspot.com), Judith Curry, Professor and Chair of the School of Earth and Atmospheric Sciences at the Georgia Institute of Technology and President (co-owner) of Climate Forecast Applications Network (CFAN) (http://judithcurry.com), Steve McIntyre (ClimateAudit.org), Steve Malloy (JunkScience.com), and Anthony Watts (WattsUpWithThat.com). For one-stop shopping of the best headlines of the day, my former committee staffer Marc Morano runs the site www.climatedepot.com. While these sites do not necessarily represent my views, I think they are wonderful vehicles that show the aggressive ongoing debate about climate science.

Of course, one of the most effective vehicles for the truth was my committee website where people could hear the viewpoints that had been silenced so long in the mainstream. My approach to an online presence was based largely on the principles put forward by conservative radio host Hugh Hewitt in his book *Blog* published in 2005. I know it seems strange today to think that my Web site, which featured a blog addressing global warming, was innovative, but at that time, not many of my colleagues on Capitol Hill were blogging, so it was actually quite a new thing. An article in the *Wall Street Journal* put it well:

> Pundits do it. Scientists do it. Even Donald Trump does it. So why shouldn't Congress blog too?
>
> As the former Chairman of the Senate Environment and Public Works Committee, Republican Jim Inhofe was a coruscating critic of climate change alarmism. Now in the minority, he plans to make sure his voice is heard over the din of the media-savvy environmental groups through a new blog [...] And their new blog is already making waves, not to mention causing some congressional tech malfunctioning.[170]

The technical malfunctioning mentioned in the *Wall Street Journal* article was due to one particular blog post where we called out Heidi Cullen of the Weather Channel for saying that meteorologists who did not believe in global warming should be stripped of their scientific certification. The Drudge Report subsequently linked to that post, and the response was so overwhelming that the entire Senate web system crashed, including all my colleagues' personal pages and committee pages.

The Hill newspaper quoted an email sent to the Senate offices by the Sergeant at Arms that day which read, "Drudgereport.com established a link on their Web site to a press release on a Senate committee Web site. This link was creating 30–50,000 queries per hour to Senate.gov, which in turn was generating a query to the Press Application for each of those hits."[171]

That was not the only time Drudge linked to our blog. It also picked up my speech on the Senate floor about the hot-and-cold media spin cycle in September 2006, which created a firestorm of support pouring into my office. My committee staff, at the time, was overwhelmed by the phone ringing off the hook from people across America, thanking us for having a refreshing voice of reason amid so much hysteria.

In thousands of emails and phone calls, Americans expressed not only their thanks but also their frustration with the major media outlets because they knew in their gut that what they had been hearing was false and misleading. Here's a brief sampling of what people were saying:

Janet of Saugus, Massachusetts: "Thank you Senator Inhofe. Finally someone with the guts to stand up and call it what it is—a sham. I think you have taken over Toby Keith's place as my favorite Oklahoman!!"

Al of Clinton, Connecticut, wrote: "It's about time someone with a loud microphone spoke up on the global warming scam. You have courage—if only this message could get into the schools where kids are being brow-beaten with the fear message almost daily."

Kevin of Jacksonville, Florida, wrote: "I'm so glad that we have leaders like you who are willing to stand up against the onslaught of liberal media, Hollywood and the foolish elected officials on this topic. Please keep up the fight!"

Steven of Phoenix, Arizona, wrote: "As a scientist, I am extremely pleased to see that there is at least one member of congress who recognizes the global warming hysteria for what it is. I am extremely impressed by the Senator's summary and wish he were running for President."

Craig of Grand Rapids, Michigan, wrote: "As a meteorologist I strongly agree with everything you said."

Dan of Westwood, Massachusetts, wrote: "This is the most concise, well researched, eloquently presented argument against Global Warming I have ever seen. Somebody in Congress has finally gotten it right!"

Adam of Salmon, Idaho, wrote: "Thank you for the brave speech made all about the hyping about alleged global warming and its causes. It took guts."

Once launched, the Inhofe Environment and Public Works Committee Press Blog quickly developed into a watchdog on news media excesses; it provided the badly needed balance that was lacking in mainstream environmental reporting. In 2007, I was honored to receive the Gold Mouse Award for having one of the best Web sites on Capitol Hill. The Gold Mouse Report and Awards were part of the "Connecting to Congress" research project, funded by a grant from the National Science Foundation. For this project CMF, partnered with researchers from the John F. Kennedy School of Government at Harvard University, University of California-Riverside, and Ohio State University to study how Members of Congress can use the Internet to improve communications with their constituents and to promote greater participation in the legislative process.

Later, in 2009, even liberal bloggers admitted that my Web site wasn't too bad. As Jonathan Hiskes of *Grist* wrote,

> Listen up, James Inhofe, because this might be the only compliment Grist ever pays you: You've got a decent website. Despite your wacked-out view that climate change is a "hoax" and your opposition to a climate bill, inhofe.senate.gov does a fair job of making your climate and energy positions clear and accessible to the Oklahomans who voted to send you to Washington. In fact, your website is more transparent than the sites of many senators who completely disagree with your views on global warming, including Democratic leaders Harry Reid (Nev.) and Richard Durbin (Ill.), along with two of the most influential senators when it comes to environmental policymaking—Barbara Boxer (Calif.) and Jeff Bingaman (N.M.).[172]

One of the best aspects of our Web site (and one of the reasons we won the Gold Mouse award) is that we had dueling headlines on the

Environment and Public Works homepage between the minority and the majority, which clearly showed that there was a strong debate on global warming. Of course, we were far more active than Senator Boxer's staff and the impact of our Web site would eventually become too much for her. She unfortunately decided to eliminate the dueling headlines and told my staff we could only post on the minority Web site, not on the homepage.

The day we shut down the Senate web system because of the Drudge link was in January 2007. To me it was an auspicious beginning to a year that, from the outside, looked as though it was going to be bad for skeptics. Democrats had taken back the Senate after the 2006 elections. I had just handed over the gavel to the new Chairman, Senator Barbara Boxer, after losing our majority. Of course, Chairman Boxer was a strong global warming advocate who promised in the early days of her Chairmanship that cap and trade was going to pass.

I may have given up the gavel, but as the Drudge incidents revealed, I knew that I had the ear of the American people, and they were hungry for the truth. As Winston Churchill said, "Truth is incontrovertible. Panic may resent it. Ignorance may deride it. Malice may destroy it, but there it is."

In many ways, the crash of the Senate Web site was a metaphor for what our message of the truth was capable of doing: exposing the global warming cap and trade hoax for what it was, and subsequently letting it all come crashing down.

4

SKEPTICISM REIGNS

I F AL GORE WAS RIGHT about one thing in 2007, it was his claim that global warming had reached a tipping point—but it was tipping in my direction, not his. In October 2007, I stood on the Senate floor and said that "future climate historians will look back at 2007, as the year global warming fears began crumbling"—and I was right.

HOLLYWOOD HYPOCRISY: GORE REFUSES TO TAKE THE PLEDGE

One of the greatest ironies of the global warming movement is that Al Gore and the Hollywood elite, who were so successful in launching the issue into the spotlight, were the very ones who ended up sealing their own campaign's doom with the American people.

They were their own worst enemies. Here they were, screaming about climate catastrophe and demanding that Americans use only one square of toilet paper, take cold showers or two-minute showers, not eat meat, and take public transportation, while clearly they were not willing to make these sacrifices themselves. In those years, it was not the science that was overwhelming but rather the hypocrisy of those who were emitting so much more carbon than the average American with their multiple mansions and private jets, telling everyone else that they had to cut back.

I wasn't going to let them get away with it. I said repeatedly that Al Gore and his Hollywood friends were happy to talk the talk but refused to walk the walk. How hard is it for these elitists to become as frugal in their energy consumption as the average American? Apparently it was impossible. They could have their limos and their private jets, but everyone else had to suffer. The American public knew they were being had.

One of the most humorous examples of this came when Al Gore and Leonardo DiCaprio teamed up to have their message "Ride mass transit" flashed on the TV screen during the 2007 Academy Awards. As Charles Krauthammer succinctly put it, "This to a conclave of Hollywood plutocrats who have not seen the inside of a subway since the moon landing and for whom mass transit means a stretch limo seating no fewer than ten."[173] Although I have to give some credit to global warming advocate John Travolta, who finally admitted, "I'm probably not the best candidate to ask about global warming because I fly jets."[174]

If John Travolta wasn't the best candidate, Gore was the worst of all. He went around saying things like: the world must embrace a "carbon-neutral lifestyle"[175] and we have just "ten years to avert a major catastrophe that could send our entire planet into a tailspin" while having a house that consumes more electricity every month than the average American household uses in an entire year.[176] And this is not to mention the multiple trips he takes on multiple carbon-emitting private jets. No, flying commercial would not do.

So Al Gore, one of the biggest carbon emitters in the world, was chosen to be the face of the campaign to eliminate carbon from our lives. This was a recipe for failure.

At the end of *An Inconvenient Truth,* the last message flashed on the screen is a challenge to America and the world: "Are you ready to change the way you live?" So I said, yes, Al Gore, are *you* ready to change the way you live? When Gore came to testify at an Environment and Public Works Committee hearing in March 2007, I asked him if he would be willing to sign a pledge—frankly, of his own making—that as a believer in catastrophic man-made global warming, would he consume no more energy at his residence than the average American household by March 21, 2008?[177] I even gave him a year to do it!

Of course, during the hearing he wouldn't answer the question and waffled around the issue, going on as he always does about how we need to transition to a clean energy future. So I said I'd be anxious to hear his answer in the questions for the record. Here's what I received months later:

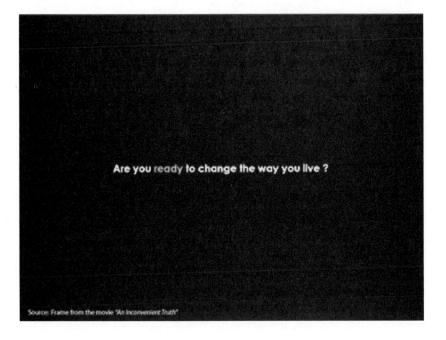

Are you ready to change the way you live ?

Source: Frame from the movie "An Inconvenient Truth"

Inhofe: As a believer that human-caused global warming is a moral, ethical, and spiritual issue affecting our survival; that home energy use is a key component of overall energy use; that reducing my fossil fuel-based home energy usage will lead to lower greenhouse gas emissions; and that leaders on moral issues should lead by example; I pledge to consume no more energy for use in my residence than the average American household by March 21, 2008.

Given that hundreds of Americans—a great many of whom could not afford offsets—would follow your example by significantly reducing their home energy consumption, will you now agree to take the pledge?

Gore: No.

DESPERATION IN THE ALARMISTS CAMPS

As one can imagine, this particular hearing caused quite a stir in the media, with every global warming reporter weighing in. The most memorable recap was aired on Keith Olbermann's show. He invited journalist Howard Fineman, who at the time was with *Newsweek* and now is a senior editor at the *Huffington Post*, to discuss what he called "the link between Inhofe and the smear campaign against Gore," which was, according to Olbermann, "venomous."[178] Their evidence was that my communications director at the time, who had once worked for Rush Limbaugh, was handing out our press release of the day, which called out Gore for his refusal to take our energy pledge. As Fineman said about my so-called "smear campaign":

> Well it is extensive. The guy you were talking about was buzzing around the press table I was sitting at, handing around press releases before, during, and after the Vice President's appearance. I think there's a few things going on here. For one, Al Gore is winning this argument, scientifically and politically.[179]

My staffer was not doing anything out of the ordinary: press staff always distribute releases and opening statements to members of the press during hearings. It is standard operating procedure. The real problem here was that I was drawing national attention to one of the movement's greatest weaknesses—its hypocrisy—so they were going overboard to compensate, practically shouting that I was losing and Gore was winning:

FINEMAN: If Inhofe is the best arguer they can come up with for their side Al Gore is in even better shape politically than he even realizes.

OLBERMAN: Yeah we have a great chance of saving the planet then. Mr. Inhofe is putting the fossil back in fossil fuels.[180]

As for Gore, Fineman called him the "'Goricle.' He's the guy everyone's listening to. He's got his answers down pat on all this thing. He goes around being a sage. He doesn't have to run for anything. He's the unofficial president of the environment now."[181]

Talk about "global governance." Their only problem was that it wasn't true: 2007 was the tipping point—the Great Goricle's prophesies of doom were wearing thin with the American people and they were coming over to my side in droves.

In the summer of 2007, Gore launched his "Live Earth" Concerts, which were held across the world—including Washington, D.C., New York, Sydney, Shanghai, and Rio de Janeiro. These performances featured "more than 150 of the world's best music acts—a mix of both legendary music acts like The Police, Genesis, Bon Jovi and Madonna with the latest headliners like Kanye West, Kelly Clarkson, Black Eyed Peas and Jack Johnson." The mission, according to the Web site, was to have twenty-four hours of music performed on seven continents in order to instill "a worldwide call to action and the solutions necessary to answer that call. Live Earth launched a multi-year campaign to drive individuals, corporations and governments to take action to solve the climate crisis."

But as MTV put it, the "results were mixed."[182] For one thing, the first snowfall in years and extremely cold temperatures were blamed for the poor turnout in Johannesburg. A Rasmussen poll published on July 8, 2007, showed that "most Americans tuned out." Only twenty-two percent of Americans even followed the concerts somewhat or very closely while seventy-five percent did not follow coverage of the event.[183] To add insult to injury, after all the time and money spent on their campaign, only twelve percent of Americans said that global warming was the most important issue to them.

Further, Rasmussen found that some of the biggest carbon emitters in the world telling everyone else to cut back wasn't really working out for them:

> Skepticism about the participants may have been a factor in creating this low level of interest. Most Americans (52%) believe the performers take part in such events because it is good for their image. Only 24% say the celebrities really believe in the cause while another 24% are not sure. One rock star who apparently shared that view is Matt Bellamy of the band Muse. Earlier in the week, he jokingly referred to Live Earth as "private jets for climate change."[184]

Americans were clearly on to them for their hypocrisy, which continued to be a persistent problem for the Live Earth campaign: for instance, Madonna, who was a key performer in these climate concerts, had just made a name for herself as one of the world's worst polluters. Having already offered the pledge to Gore, I thought why not extend it to other eco-hypocritical celebrities as well? In April 2007, in the spirit of Earth Day, I challenged global warming activists/celebrities such as Laurie David, John Travolta, Leonardo DiCaprio, and Madonna to do what former Vice President Al Gore refused to do—live up to their environmental rhetoric by reducing their home energy usage to the level of the average American household by Earth Day 2008.[185]

I thought it was the least they could do since they were so worried that mankind only had ten years left to act in order to avoid a climate catastrophe, and they had made personal energy use a cornerstone of their pleas to the general public to save the planet. I have yet to get one of them to sign my pledge.

THOSE PESKY BLIZZARDS: GLOBAL WARMING OR CLIMATE CHANGE?

The alarmists also had another serious problem: the weather didn't always cooperate with their predictions of tropical doom—they didn't know what to say when it turned cold. With a dire future in store of melting ice caps, drowning polar bears, and rising sea levels, Al Gore and the media had built up such a strong narrative that global warming was leading to higher temperatures, any cold spell threw them for a loop. Keith Olbermann's clip on Gore's visit to Capitol Hill provides a good example of the kind of rhetoric that was always getting them into trouble:

> **OLBERMAN:** What's Inhofe's story? ...Does he explain the oil companies who are funding some of the garbage science that says oh no everything's great. Don't anybody panic. That sweat on your brow in the middle of February has nothing to do with global warming.[186]

But the problem was that places across America were not experiencing sweat on their brows in the middle of February but, instead, record cold temperatures—and that was kind of inconvenient for them. At the March 2007 hearing, I challenged Gore on that point. He had mentioned that the fires that were occurring in Oklahoma that year were due to global warming. So I asked him, "How come you guys never seem to notice it when it gets cold?"[187] I held up a document from the National Oceanic and Atmospheric Administration that showed that there were 183 record cold temperatures in January 2007—183 of them! As for Oklahoma, we had three days that year that were the coldest days in history. So I asked: where is global warming when you really need it?

That was a question that continued to haunt the movement. Their heated rhetoric failed every time the snow fell: they needed to modify it to accommodate those inconvenient blizzards and record cold temperatures. Al Gore later went on to claim that heavy snowfalls are completely consistent with man-made catastrophic global warming, saying in a February 27, 2010 *New York Times* op-ed, "Just as it's important not to miss the forest for the trees, neither should we miss the climate for the snowstorm." The American people were suspicious to say the least, especially when they noticed a deliberate shift in terms over this period from catastrophic "global warming" to "climate change."[188]

AL GORE'S NEW HOME

"Snowmageddon"—the intense blizzard that shut down Washington, D.C., for weeks in February 2010 was what really spurred a crisis of messaging in the global warming camps.

Because of the blizzard, the airport was closed, all the federal build-

My daughter Molly's family at the igloo

ings were closed, and very few people were even leaving their houses. Just days before the storm struck, my daughter Molly and her family arrived in Washington for an annual event that I sponsor called the African Dinner. This year was special because my beautiful granddaughter Zegita Marie was going to speak at the dinner. You see, we found Zegita in Ethiopia as a baby—she had been left an orphan. Molly and her husband already had three boys and always wanted a girl. They instantly loved Zegita Marie, as we all did, and decided to adopt her.

God may have been having a little fun with the alarmists that week too: the scheduled Environment and Public Works Committee hearing on "Global Warming Impacts, Including Public Health, in the United States" had to be postponed because the blizzard had essentially shut the government down. Of course, the logic of the alarmists was that the global warming hearing was cancelled due to a blizzard that was caused by global warming. Senator John Kerry reminded us that we were still facing the grave threat of global warming and that the "solution" was still on the horizon: as he said in an article for *The Hill*, "those who think climate change legislation is dead for the year are 'dead wrong.' Those who think blizzards and record snow falls in Washington will make it tough to move a global warming bill are guilty of 'inside the beltway' thinking."[189]

Molly and her husband and kids decided to make the most of the snow and built an elaborate igloo right outside the Library of Congress—you could put four people in there. Then they put on the finishing touch: a sign that read, "Al Gore's New Home—Honk if you love global warming." So as you can see, skepticism runs in our family.

They sent me a picture and I thought it was so clever I put it up on my Facebook page right away. The media blizzard that ensued was really remarkable. But of all the press, Keith Olbermann, who had just a few years earlier said we'd be wiping sweat off our brow in the middle of February, was the one who took it the hardest. He featured Molly's family on his show that night as "The Worst Family in America." I guess some people just can't take a joke.

I'm a public figure, an elected official, and being a target is part of the job. It's quite another thing for my family to be subjected to ridicule and vicious name-calling. But I'll let my daughter Molly speak to that:

How An Afternoon of Family Fun Ended Up in Global Warming Headlines
—By Molly Inhofe Rapert, daughter of U.S. Senator James M. Inhofe

It's hard to imagine that choosing to spend a family afternoon in the outdoors together could result in a polarizing political event, but that's exactly what happened in February of 2010. It all began with a special trip to Washington, D.C. to watch our youngest child, Marie, speak at an African event associated with the 2010 National Prayer Breakfast. Our family of six lives in Fayetteville, Arkansas and includes my husband, Jimmy, and our children, Jase, Luke, Jonah, and Zegita Marie. At the time of this event, the kids were 14, 12, 11, and 9. Marie was born in Addis Ababa, Ethiopia, in 2001 and beautifully spoke of her adoption journey, and love for Africa, on Thursday evening, February 6.

The weather took a turn for the worse that evening and, by Friday morning, had developed into a record-breaking snowstorm. My husband, Jimmy, always finds great things for our family to do outdoors—hiking, camping, and more. So he saw this as the perfect opportunity to spend an afternoon outside, building an igloo near the

Library of Congress. Five hours later, an impressive six-foot igloo had emerged with Capitol Hill as the backdrop.

Our oldest son, Jase, added a touch of humor with a sign reading "Al Gore's New Home" with the flip side: "Honk if You Love Global Warming." Jase had completed two science fair projects that examined the media perceptions of global warming/cooling and the projects just sparked a natural connection to the irony of being in Washington, D.C. in a historic storm, steps away from where Vice-President Gore had worked on his book, *An Inconvenient Truth*.

Jase's sign went up on top of the igloo, car horns honked, and folks stopped to have their photo taken. It all seemed harmless enough— actually, not just harmless...I'll admit that, as parents, we were proud that we had managed to get our kids out of the house, spending 5 hours playing outside together on a very cold day.

Washington, D.C. was virtually shut down, the Smithsonians were closed, we were simply making the best out of the situation at hand. We enjoyed visiting with people as they stopped to look at the igloo, and everyone that stopped was light-hearted and happy, whether they believed in man-made global warming or not. It was a great day.

We call my dad PopI (short for Pop-Inhofe). PopI loaded the photo of the igloo on his Facebook page, where it was picked up by *Roll Call* then *Heard on the Hill*. From there, the photo snowballed with the same ferocity of the snowstorm. The initial coverage adopted the same tone as the kind-hearted people that had stopped to talk to us on the street— it was a funny quip and everyone understood that it was simply a family outing on a snowy day, with no intense political agenda intended. We headed back to Fayetteville, with great memories of our snowy trip to Washington, D.C. Then Keith Olbermann entered our lives.

A common end-of-show segment on the Olbermann show is to identify the worst person in the world. That Sunday evening, Mr. Olbermann, in a bizarre twist, far removed from what anyone could deem true journalism, decided to name our family one of his worst families in the world for making a joke about climate change "in a storm that killed people." Actually no one was killed by the storm in Washington, D.C., but that's only one of the many journalistic flaws he commits on a regular basis. He didn't even bother to pronounce our names correctly.

We did not see his show when it aired. Our wonderful pediatrician, a staunch conservative who watches Olbermann to "see what the enemy is saying," texted us at midnight to say she had never been so appalled and that we had to watch it. After viewing it together, Jimmy and I knew we needed to show it to the kids before they went to school, in case any of their friends had seen it.

We gathered around our kitchen table. The look on the younger kids' faces was hard to watch. How confusing this must have been to them—we had worked hard to teach them to respect authority and to encourage them to watch the news, learn about the world around them. For them to see an adult speaking about their family in such an aggressive, abrupt, highly inappropriate way was insulting. Their faces revealed their confusion.

You can imagine that, from their point of view, they had spent a day in the snow building an igloo—no political agenda, no political discussion. Just a lot of good, old-fashioned work making lots and lots of blocks of snow. They couldn't quite equate their igloo experience with the man on the computer screen ranting and raving about them by name. To say they were "taken aback" perhaps describes it best. We tried our best to use it as an opportunity to explain the importance of always conveying the truth of a situation, illustrating how Mr. Olbermann had used the situation out of context to further his own agenda. We explained that they had done nothing wrong; it was a lighthearted day that someone else was attempting to use to his advantage. Our oldest son, Jase, lightened up the situation when he said "Worst family in the world? From Olbermann? I wear that as a badge of honor."

Olbermann's show sparked remarkable controversy in the media. We were shocked with the intensity of the comments on the Internet as well as the language that was used. Some bloggers harshly criticized us, saying that we asked for this by mocking the vice-president. I would argue two things. First, many people make light-hearted jokes about global warming whenever the weather turns cold...we simply put it on paper. We didn't shout it from the mountaintops, we didn't set out with a political agenda, we just made an igloo for fun and then put the sign up as a minor afterthought at the end of an enjoyable day. Second, I don't know of any reasonable human being who could possibly equate a sign made by a fourteen-year-old with the vitriol of Keith Olbermann—an adult who chose to identify a family on national television in a diatribe. How Olbermann and his staff could look at a photo of four young children having fun outdoors and, with malice, go after them in an attempt to elevate themselves—it is hard for me to fathom.

Others were not amused with Olbermann's attack. Rush Limbaugh and Sean Hannity both aired segments commenting on the igloo situation and commending our family for building it. Blogger Amy Ridenour levied a harsh criticism at Olbermann's sponsor, stating "GE should be ashamed of itself for allowing its personnel to attack children on air. These kids probably are sophisticated enough to realize Olbermann's just doing it for attention, but it's still pathetic to see a giant corporation going after kids." Individual bloggers rallied to our defense, hammering both Olbermann and the bloggers who had joined his side. But the energy gathered on both ends of the spectrum. One of my personal favorites came from blogger, They Gave Us a Republic, who stated on his website that I suffered from "congenital, multigenerational stupidity." The Progressive Electorate blog claimed that PopI had "used" his grandchildren to make a political point, making us build an igloo which mocked a snowstorm that killed people. Fortunately, this level of stupidity didn't need to be defended as most Americans recognize that you don't need to force kids to play in the snow—although perhaps that blogger had a childhood where he preferred to sit in front of a computer rather than play outside? Oh yes, and Grant Lawrence's blogsite called our children "mentally handicapped" spawns who are academically inferior to a damp sponge. The poor fellow would be so disappointed to learn that our kids are actually all honor roll students. The wonderful blog, Maggie's Notebook, jumped to our defense, pointing out the absurdity of these and other statements.

By the end of the week, the igloo story had been picked up by a variety of news sources including: ABC Nightly News, CNN, Fox News, *Good Morning America*, *Neil Cavuto Business*, *The Washington Post*, *New York Times*, CBN, CBS, CNS, NBC, *Newsmax*, Rush Limbaugh, Sean Hannity, and *Wall Street Journal*.

I'll admit that as a marketing professor, if I ever hoped to be quoted in the *Wall Street Journal*, I never expected it to be for an igloo! Here is a sampling of the headlines from the various blogs and internet sources, illustrating the extreme takes on the story:

- Inhofe's grandchildren build igloo to mock killer snow-storm[190]

- Jim Inhofe, America's Worse Senator, also has America's Worse Grandchildren[191]

- Olbermann names Inhofe family worst person in the world

- GOP Sen Inhofe's global warming denial a family affair

- Global warming denier Sen. James Inhofe's family builds a very special igloo for Al Gore

- Inhofe's Al Gore Igloo

- Inhofe family talks about igloo, national media coverage

- Inhofe family mocks Al Gore's global warming

- Al Gore's new house on Capitol Hill

- Global warming snow job

- Family's igloo draws debate over warming

- Olbermann Inhofe family worst family in world: igloo frosts critics

- 14-year-old Jase Rapert tells off Olbermann

- Inhofe igloo frosts critics

- DC igloo takes aim at global warming

- UA professor's ice storm story continues to snowball

- Al Gore's new home don't you dare

- Family in hot seat for global warming joke

As Tina Korbe stated in an article written for the University of Arkansas Traveler, "It's difficult to say which was bigger: the historic ice storm that prompted UA marketing professor Molly Rapert and her family to build an igloo on Capitol Hill—or the media blizzard the igloo inspired."[192] This was an interesting lesson for our family, even more relevant to society today as we watch how quickly internet sources and pundits will jump on a story and stretch it beyond the limits of the original situation. On one hand, it is inspiring that so many people care about issues and want to be involved. Whether they ardently believe that government intervention is needed to help pro-tect the earth—or whether they believe the earth is naturally moving through warming and cooling cycles and that it is foolhardy to pursue government-based initiatives. On the other hand, it is disheartening to witness the willingness of people to publicly express such extreme, intense opinions that are not fact-based. We were grateful for bloggers, such as Amy Ridenour and Maggie's Notebook, who were willing to use their blog forums to counter comments that stepped beyond the bounds of decency.

An enjoyable family afternoon together; a five-hour-in-the-making six-foot igloo; and a funny quip that captured the sentiments of many make up the great igloo saga of 2010. If that makes us the worst family in the world, I'll join Jase and wear that badge with honor.

....

Olbermann, by the way, left the network not long after that.

My daughter Molly is a brilliant marketing professor with a PhD from the University of Memphis. She received the 2010 Hormel Master Marketing Teacher Award and was an Inaugural Recipient of the 2011 University of Arkansas Honors College Distinguished Faculty Award. She is also the recipient of the Beta Gamma Sigma Outstanding Teaching Award (2007), the Sam M. Walton College of Business Excellence in Service Award (2006 and 1993), the Arkansas Alumni Association Distinguished Faculty Achievement in Teaching Award (2002), the Walton College of Business Excellence in Teaching Award (2001 and 1998), and the Walton College Excellence in Advising Award (1996).

To my great surprise, one member of the liberal establishment media, Dana Milbank of *The Washington Post*—one who had his own fun at my expense several times—understood the significance of the igloo for his side's argument. As he wrote:

Still, there's some rough justice in the conservatives' cheap shots. In Washington's blizzards, the greens were hoist by their own petard. For years, climate-change activists have argued by anecdote to make their case. Gore, in his famous slide shows, ties human-caused global warming to increasing hurricanes, tornadoes, floods, drought and the spread of mosquitoes, pine beetles and disease. It's not that Gore is wrong about these things. The problem is that his storm stories have conditioned people to expect an endless worldwide heat wave, when in fact the changes so far are subtle. Other environmentalists have undermined the cause with claims bordering on the outlandish; they've blamed global warming for shrinking sheep in Scotland, more shark and cougar attacks, genetic changes in squirrels, an increase in kidney stones and even the crash of Air France Flight 447. When climate activists make the dubious claim, as a Canadian environmental group did, that global warming is to blame for the lack of snow at the Winter Olympics in Vancouver, then they invite similarly specious conclusions about Washington's snow—such as the Virginia GOP ad urging people to call two Democratic congressmen "and tell them how much global warming you get this weekend." Argument-by-anecdote isn't working. Consider the words of Sen. Jeff Bingaman (D-N.M.), chairman of the energy committee, who told *The Hill* newspaper last week that the snow "makes it more challenging" to make the case about global warming's danger to people who aren't "taking time to review the scientific arguments."[193]

So he was saying that my family had a point. Alarmists were always going on about how the overheating planet is causing every weather disaster known to mankind, but there was always a crisis of messaging when it came to extremely cold temperatures. Milbank concluded his column writing, "If the Washington snows persuade the greens to put away the slides of polar bears and pine beetles and to keep the focus on national security and jobs, it will have been worth the shoveling."[194]

ABC NEWS: EGG ON THEIR FACE

Not everyone in the media received Milbank's memo about not using weather to justify global warming policy.

It was a very hot summer day in July 2010 when Jon Karl with *ABC News* approached my office to ask if I would give an interview on the prospects of global warming policy—outside in the heat. Their intentions were obvious, so my staff asked me if I wanted to do it. Of course I did—I enjoy ambushes—so we headed outside, ready to turn the tables on them.[195]

Sure enough, instead of the usual one cameraman and one reporter, there was a large production crew set up, and a crowd had gathered outside to watch, so I knew they were planning to do something big. Karl began with questions on the prospects of cap and trade but then shifted quickly to what he really wanted to ask me about: the weather. It was so predictable. "How's that igloo doing now?" he asked. Here we go again.

Then Karl had brought out his incontrovertible proof that global warming was happening: a pan with an egg in it. They thought it was so hot outside that the egg would fry—only it didn't. After the botched ambush, the crew tossed the egg out on to the grass. When my communications director started to take pictures of the unfried proof of global warming, the ABC news crew came sprinting back to clear the evidence. The crowd that had gathered was given quite a show, but not the one they expected.

THE "CONSENSUS" BEGINS TO COLLAPSE

In March 2007, the IPCC fourth assessment boldly declared that man's contribution to global warming was "unequivocal," Hollywood and the mainstream media had been peddling the scientific "consensus" with a vengeance, and yet, a *Los Angeles Times*/Bloomberg poll in August 2006 found that most Americans did not attribute the cause of any recent severe weather events to global warming, and the portion of Americans who believed that climate change is due to natural variability has increased over 50 percent in the last five years. After all that money and effort, this must have been quite a blow.[196]

If Hollywood hypocrisy had cooled the public's reception to the alarmists, their excessive hysteria, which inflated the issue beyond their wildest dreams, was exactly what propelled so many scientists, even those from the left wing, to come forward in dissent. Scientists began to feel that their work was being pushed to the side while activists made the most outlandish claims; many became increasingly uncomfortable with the alarmists' aggressive agenda. As a result, several key scientists made the "conversion" from a believer to a skeptic.

The biggest shock to the global warming camp was probably the conversion of Dr. Claude Allegre—a renowned French geophysicist, a former French Socialist Party leader, a member of both the French and U.S. Academies of Science, and one of the first scientists to sound the global warming alarm—who changed around 2006 from being a believer to a skeptic. This was a guy who marched up and down the street twenty years ago saying man-made gases are going to bring the world to an end. Now he was saying that the cause of climate change is unknown and even accused the climate alarmists of being motivated by money. In a September 2006 article, he said, "The ecology of helpless protesting has become a very lucrative business for some people!"[197] I thought that it was so ironic that a free market conservative capitalist in the U.S. Senate and a French Socialist scientist were both saying that sound science is not what is driving this debate, but greed by those who would use this issue to line their own pockets.

Allegre was not the only prominent scientist to convert. Astrophysicist Dr. Nir Shaviv, a sharp young astrophysicist from Israel, also recanted his belief that man-made emissions were driving climate change, and said that solar activity can explain a large part of warming in the twentieth century. As he said in a February 2, 2007, *Canadian National Post* article:

> Like many others, I was personally sure that CO_2 is the bad culprit in the story of global warming. But after carefully digging into the evidence, I realized that things are far more complicated than the story sold to us by many climate scientists or the stories regurgitated by the media. In fact, there is much more than meets the eye.[198]

Botanist Dr. David Bellamy, a famed UK environmental campaigner, former lecturer at Durham University, and host of a popular UK TV series on wildlife, also converted into a skeptic after reviewing the science. Bellamy said that "global warming is largely a natural phenomenon" and said that catastrophic fears were "poppycock." "The world is wasting stupendous amounts of money on trying to fix something that can't be fixed," and "climate-change people have no proof for their claims. They have computer models which do not prove anything."[199] Bellamy paid a steep price for his conversion: he was derided by many environmental groups, and they severed their ties with him despite his long record of environmental activism, which included being arrested while trying to prevent loggers from cutting down a rainforest in Tasmania, saving peat bogs, and other endangered habitats.

Finally, meteorologist Dr. Reid Bryson, the founding chairman of the Department of Meteorology at the University of Wisconsin (now the Department of Oceanic and Atmospheric Sciences), who was a key figure promoting the ice age scare of the 1970s, also converted into a leading global warming skeptic. In a May 2007 issue of *Wisconsin Energy Cooperative News*, he said,

> Before there were enough people to make any difference at all, two million years ago, nobody was changing the climate, yet the climate was changing, okay? All this argument is the temperature going up or not, it's absurd. Of course it's going up. It has gone up since the early 1800s, before the Industrial Revolution, because we're coming out of the Little Ice Age, not because we're putting more carbon dioxide into the air.[200]

Another remarkable thing happened just a year before. On April 11, 2006, sixty prominent scientists, many of whom advised the Canadian Prime Minister in the 1990s to ratify Kyoto, wrote in an open letter to Prime Minister Stephen Harper saying,

> Significant [scientific] advances have been made since the [Kyoto] protocol was created, many of which are taking us away from a concern about increasing greenhouse gases. If, back in the mid-1990s, we knew what we know today about climate, Kyoto would almost certainly not exist, because we would have concluded it was not necessary.[201]

Also important was that the Czech President Vaclav Klaus said on February 8, 2007, that fears of catastrophic man-made global warming were a "myth" and critiqued the UN IPCC process, calling it a "political body." He also remarkably said that other government leaders would also speak out if it were not for the fact that "political correctness strangles their voice."[202]

While major media news outlets continued to ignore the prominent scientists and leaders such as Allegre, Shaviv, Bellamy, Bryson, Czech President Klaus, and the sixty Canadian scientists who were becoming increasingly skeptical, I noticed that smaller news outlets across the country were actually doing their job and providing balanced coverage of global warming, and in the process, they were uncovering hundreds of skeptical scientists.

Award-winning Chief Meteorologist James Spann of an Alabama ABC TV affiliate, who holds the highest level of certification from the American

Meteorological Society, wrote in a January 18, 2007, blog post: "I do not know of a single TV meteorologist who buys into the man-made global warming hype. I know there must be a few out there, but I can't find them."[203] This fascinated me, so I asked my staff to begin watching for and compiling news clips that mention scientists who disputed man-made global warming catastrophe. James Spann has the distinction of being the first skeptic scientist to be included in a list that grew to over 450 in 2007; it reached 750 by 2009, and today my former communications director, Marc Morano, keeps the list up to date. This list now stands at well over 1,000 scientists.[204]

"ELECTIONS HAVE CONSEQUENCES": MY FRIEND, BARBARA BOXER

Most people are surprised when I say this, but I really like Barbara Boxer. She's always wrong on the environment, and our debates have been very fierce at times, but we've been great friends for many years, and we have a lot of lively memories as we've alternated the leadership of the Environment and Public Works Committee. I handed the Chairmanship over to her in January 2007, and, as she put it at that particularly fiery hearing with Al Gore, waving the gavel in front of my face, "You don't do this anymore. Elections have consequences." It was all in good fun, but I joked that she's just set herself up for a repeat of those words when things change back again.

So she was making the rules in 2007 and the structure of the committee was very different than it was when I held the gavel. For one thing, instead of examining the economic implications of the "solutions" to global warming, as I was determined to do as Chairman, we had twenty hearings focusing on the dangers of global warming, including "Global Warming and Wildfire," "The Examination of the Views of Religious Organizations Regarding Global Warming," and "Global Warming Impacts on the Chesapeake Bay"—my favorite being the affect of global warming on the tourism industry.

The irony seemed to be lost on my colleagues that we were having endless discussions about how urgent global warming was and how we must act now or the world will end; yet they were in no hurry to discuss the "solution" of cap and trade. It wasn't until October 2007 that Senator Boxer finally held a hearing on the newest cap and trade bill introduced by Senators Joseph Lieberman and John Warner.

In truth, I wasn't at all surprised that there was very little effort to examine cap and trade seriously. No one on their side wanted to discuss the economic pain that it would cause. It was legislation that would, as the sponsor of the bill admitted, cost "hundreds of billions of dollars," and, as *The Washington Post* put it, "require a wholesale transformation of the nation's economy and society."[205] I could see why my good friend and climate foe didn't want to broadcast that this bill would be largest tax increase in American history.

Barbara and I had a big interview with Larry King on January 31, 2007, just a few weeks after she had taken up the gavel. I was waiting in the greenroom where they mic you up a few minutes before you go on air, when Barbara came out of the make-up room ready to battle it out with me on cap and trade. She had an entourage of staff with her so I turned to them: "Don't worry, it won't get too hostile. You may not believe it but Barbara and I actually like each other," I said, patting her on the head. But I apparently destroyed her hairdo and her staffers were horrified; she immediately ran back into the make-up room with her staffers close at her heels to get it fixed.

Barbara and I have fought each other on cap and trade and global warming for many years, but we've had some good laughs too. At one memorable hearing, I gave her a gift: it was a coffee mug with a picture of the world's coastline being inundated with water from global warming when one pours liquid in the cup. But she had a good sense of humor about it: as seriously as she takes the issue, she couldn't help but laugh.

Larry King put it well when he said at the close of our interview with him, "Senator Barbara Boxer, Democrat of California. James Inhofe, Republican of Oklahoma. And, by the way, they are good friends."[206]

CAP AND TRADE AGAIN: LIEBERMAN-WARNER DIES

In October 2007, Al Gore and the IPCC together won the Nobel Peace Prize for their work to bring widespread attention to the emergency of global warming. Yet even so, and despite the efforts of the Hollywood elite and the mainstream media, Senator Boxer still couldn't sell the Lieberman-Warner cap and trade bill in the Senate.

Here's how it all shook down: the United Nations Climate Change Conference in Bali was just around the corner, so Barbara was in a rush to get Lieberman-Warner passed out of Committee so she could meet world leaders with a climate "victory" under her belt. Along with Senators Voinovich and Barrasso, I requested a full economic analysis of the bill by the Environmental Protection Agency and the Energy Information Administration before proceeding to a committee vote. We maintained that it was irresponsible to move forward without knowing the full extent of the economic damage resulting from this bill. There were also many attempts by Republicans to offer amendments that would mitigate its negative economic consequences—but these were completely rejected and only Democrat amendments were added.

The business meeting to pass the bill was in December 2007. It was the first time a cap and trade bill ever made it through the Environment and Public Works Committee and all Republicans, except for Senator Warner, voted against it. We had been ignored; in her rush to get it out the door, Barbara opted to wait to address the more significant obstacles, including the economic damage of the bill, until it reached the Senate floor.

The "obstacles" were indeed significant. At the time, Duke Energy Corp. Chairman Jim Rogers warned that the bill will cause a "customer revolt" due to a rise in electricity bills by as much as 53 percent in 2012.[207] Additionally, the widely respected nonpartisan Charles River Associates issued a November 8 analysis of the bill, revealing it would result in $4 trillion to $6 trillion in welfare costs over forty years and up to $1 trillion per year by 2050.[208] Even the co-author of the bill, Senator Lieberman, conceded on November 1 that his bill would cost "hundreds of billions of dollars." The American Council for Capital Formation's analysis on November 8 found the bill would lead "to higher energy prices, lost jobs and reduced [gross domestic product]."[209] Two unlikely groups agreed on this one point as well: the AFL-CIO was worried, saying that "the bill would cost jobs by giving a competitive advantage to foreign companies that aren't subject to similar restrictions,"[210] and the U.S. Chamber of Commerce said that it "does not adequately preserve American jobs and the domestic economy."[211]

Not only would Lieberman-Warner have been a disaster for the economy, it also would have brought our country to an abrupt halt. That's because the bill did not include nuclear, and without it, we would have been severely lacking in baseload power, as fossil fuels were gradually phased out. The lack of nuclear may have been a blessing in disguise for my good friend John McCain, as it gave him an out not to support a bill that was very similar to the one he introduced a few years before. He had already launched his campaign for president and, understandably, he did not want to vote for the largest tax increase in American history.

Senator Reid brought Lieberman-Warner to the Senate floor in June 2008, but the results were already in before the vote was even taken. It was obvious that the Democrats were not at all serious about the bill. To put it in perspective, the 1990 Clean Air Act amendments were considered on the Senate floor for five weeks, due to the substantial impact they would have. Lieberman-Warner was even more consequential economically, yet Democrats had no interest even in considering our amendments that would protect American families and workers from the devastating effects of this bill. On June 3, *Roll Call* quoted frustrated Democrat staffers saying, "We have no strategy, no message, and no plan" and, "Boxer is walking us off a cliff."[212]

During the floor debate all the Democrats, true to form, were talking about nothing but the science and how global warming was real. Senator Boxer was no exception—in fact, she said very little about the bill itself; it was all about how we must act now because the science was settled. So I went down to the Senate floor and conceded the science. I said, "Let's say we are headed toward unspeakable catastrophe, would cap and trade save us?" Absolutely not. That's because only America would be taking action: China, Mexico, and India have no intention of inflicting this kind of economic harm on themselves so as jobs move to those countries where they don't have any emissions limits, global temperatures would actually increase. It would be all economic pain—significantly higher energy costs, hundreds of thousands of lost jobs, and a depressed economy—for nothing.

This strategy really threw Senators Boxer, Kerry, and others for a loop as they came to the Senate floor fully prepared to have a science debate with their doom and gloom pictures of polar bears, melting glaciers and hurricanes. Instead, what they got was a debate on economics and costs, which until then had taken a back seat to the dramatic images.

During that week in June when the Lieberman-Warner bill was debated on the Senate floor, I carried around a memo with four themes based solely on economics. They were taxes, jobs, gas prices, and nuclear power. The themes quickly got the attention of the American people and my fellow Republican colleagues, setting the stage for the economic debate on the Waxman-Markey bill later in 2009 and 2010.

The news hit them hard: "Government studies confirm this bill will only raise gas prices";[213] "$6.7 trillion in the form of higher gasoline and electricity bills";[214] "1.8 million jobs lost by 2020 and 4 million by 2030 according to the National Association of Manufacturers."[215] The economic impacts of the bill drove the debate and we continued to build on these themes as the week went on.

I argued how many American jobs would be lost to developing countries such as China, India, and Brazil that have been allowed to emit greenhouse gasses without international criticism. For instance, China had built 117 government-approved coal-fired power plants in 2005— a rate of roughly one every three days, according to official figures. Without international participation, I argued, which this bill failed to address adequately, global concentrations of greenhouse gases will continue to increase, even if America were to nearly eliminate its emissions. This news was indeed alarming to my opponents, and it continued to throw them off as they hastily tried to respond to the economic findings with continued emphasis on science and doom and gloom scenarios.

It wasn't just industry making these assumptions. The independent Energy Information Administration said the bill would result in a 9.5 percent drop in manufacturing output and higher energy costs, and that it would be worse unless we could build 268 new nuclear plants by 2030.[216] This country had already lost 3 million manufacturing jobs since 2000. The EPA itself even estimated that the Lieberman-Warner bill would increase fuel costs an additional 53 cents per gallon by 2030 and $1.40 by 2050.[217]

Sponsors of the bill also failed to tell the American people that it would generate over $6.7 trillion in revenues through 2050 from the sale and auction of carbon allowance to energy users.[218] Unfortunately, that $6.7 trillion cost would be passed on to families and workers across the country in the form of higher gas prices, higher electricity bills, more expensive consumer goods, and higher workplace costs.[219] In fact, new funding for government programs, minus any set asides for transition assistance or tax relief to states, industry, or consumers under the Act is a staggering $4.2 trillion. I made sure to shine light on these numbers, emphasizing what it would mean for American families. For example, according to various economic models, this bill would cost $3,298 to Oklahoma families in 2020 and cut over 51,000 jobs through the life of the bill.[220]

We continued to hammer these themes during that week, but even though my staff suggested that I stick to the economics during the debate afterwards I still had a few things to say about the science so unbeknownst to them, I went back down to the floor to give them an extra dose of reality on that point, too.

Senator Boxer maintained Democrats "had 54 Senators come down on the side of tackling this crucial issue now" following the cloture vote of 48–36 which effectively killed the bill. But she did not take into account a letter that was signed on June 6 by ten Democratic Senators who explicitly stated that "they cannot support final passage" of the "Climate Tax Bill." The letter showed that Boxer would have only had at most 39 votes to support final passage of the bill. As the letter stated,

> As Democrats from regions of the country that will be most imme-
> diately affected by climate legislation, we want to share our concerns
> with the bill that is currently before the Senate. We commend your
> leadership in attempting to address one of the most significant threats
> to this and future generations; however, we cannot support final pas-
> sage of the Boxer Substitute in its current form.[221]

The ten Democratic Senators were: Debbie Stabenow (D-MI), Senator John D. Rockefeller (D-WV), Carl Levin (D-MI), Blanche Lincoln (D-AR), Mark Pryor (D-AR), Jim Webb (D-VA), Evan Bayh (D-IN), Claire McCaskill (D-MO), Sherrod Brown (D-OH), and Ben Nelson (D-NE). Of the ten Senators, only Senator Brown voted against cloture.

All in all, the bill didn't have the slightest chance of passing. Republicans were prepared to debate the bill and were ready to offer amendments. But the Democrats did not want to debate, much less vote, on our amendments that were aimed at protecting American families and workers from the severe economic impacts of this bill. Even if it made it through the Senate, Democrats knew the House would never pass it, and even if they jumped that hurdle, Bush would have vetoed it. But that wasn't really the point. The point was to show leaders at the UN Climate Conference that the Senate had taken action. They had a symbolic victory just in time for Bali and their plan was to use it as a model for future success—but it would never turn into an actual victory.

As I predicted, it all came crashing down when the economic reality of the bill was exposed. When faced with the inconvenient truth of the bill's impact on skyrocketing energy prices, very few Senators were able to come out in support of the bill.

I was surprised that environmentalists were so "stunned that their global warming agenda was in collapse" in the wake of the bill's failure because that was obvious to me. That quote came from a June 6, 2008, article in the *Wall Street Journal* which also noted that "green groups now look as politically intimidating as the skinny kid on the beach who gets sand kicked in his face. Those groups spent millions advertising and lobbying to push the cap and trade bill through the Senate."[222]

THE GORE EFFECT

God has a sense of humor. December 5, 2007, was a particularly cold and snowy day to be passing the Lieberman-Warner global warming bill out of committee. In the middle of the markup, just before I voted no on Lieberman-Warner, I walked outside the Dirksen Senate Office Building to meet a reporter from the Business and Media Institute who was braving the cold to catch me for an interview. I told him that the bill would suffer the same fate as the McCain-Lieberman bill on the Senate floor due to its enormous pricetag. Looking around at the snowflakes, I asked the question I asked Gore in that fateful hearing: Where's global warming when you need it?

5

THE MOMENT ARRIVES AND THE MOVEMENT COLLAPSES

THE MOUNTAIN

On December 3, 2009, just days before the Copenhagen climate conference, Rachel Maddow aired a five-minute segment featuring me, the "unmovable" denier:

> **ANNOUNCER:** For climate change activists, James Mountain Inhofe is an inconvenient truth [...] The M. stands for Mountain. No really his name is James Mountain Inhofe and what a mighty unyielding alp he is. All others are just foothills.

[…]

ANNOUNCER: Yes, elections do have consequences. Senator Inhofe just won reelection last year. So we'll be in the shadow of this mountain until at least January 3, 2015.

RACHEL: Two questions for you Kent. Number one: his middle name is *actually* Mountain?

KENT JONES: Yeah.

RACHEL: Not a TMI invention?

KENT: For real. Mountain. James Mountain Inhofe.

RACHEL: Did he also really suggest that the Weather Channel was trying to boost its ratings?

KENT: He said they'd like that. Yeah, that'd be great. They'd love it if we were afraid all the time.[223]

Rachel's segment was one of the last major efforts to go after me just days before I landed in Copenhagen and declared vindication, but as I said at a bloggers' luncheon at the Heritage Foundation when they asked me what I thought about the clip, "You know, I've really grown to like that gal. She thinks she's saying such hateful things about me, but they're all true"[224]—including the Mountain part. Mountain is my mother's maiden name. If I was indeed a "mountain of indignation" for global warming activists, as Rachel's segment claimed, it was only because I was a vehicle for the truth and that was an insurmountable obstacle for them. For all the time, money, and effort that they poured into their message that global warming was man-made and catastrophic, Americans were starting to see the truth: the science was not settled and their "solutions" were dead on arrival.

THE STARS ALIGN

In January 2009, from the outside looking in, it seemed that the United States was poised to open a new chapter in environmental policy. Barack Obama—the climate savior who had promised to slow the rise of the oceans and heal the planet under his plan of a cap and trade system—had just been sworn in as President of the United States. His allies in Congress were ready to take action: Speaker of the House Nancy Pelosi and Senate Majority Leader Harry Reid had made commitments that they were going to pass cap and trade. The international stage was set for President Obama to arrive in Copenhagen in December and finally bind the United States to an international agreement to limit greenhouse gases. The stars were aligned for the first time ever. The moment they had been waiting for had finally arrived.

But by the end of 2009, it was all over, leaving the *UK Guardian* to ask bewildered:

> How can everything have gone so wrong so quickly? A year ago, the prospects for successful climate change regulation were bright: a new U.S. president promised positive re-engagement with the international community on the issue, civil society everywhere was enthusiastically mobilising to demand that world leaders 'seal the deal' at Copenhagen, and the climate denial crowd had been reduced to an embarrassing rump lurking in the darker corners of the internet. Now there seems to have been a complete reversal.[225]

In January of that year, while my environmental friends were popping the champagne, blinded by the euphoria of being so close to victory, I went to the Senate floor to give them a dose of reality. I said that no matter how closely stars were aligned on Capitol Hill and Pennsylvania Avenue, cap and trade would never pass and the science would implode under the pressure of its own flaws.

MEDIA MANIA CONCEDES TO SKEPTICISM

In fact, the irony is that the movement fell from inevitability to failure precisely because the stars were so aligned. With the full control of both Houses of Congress and the White House, many believed that cap and trade *could* become the law of the land, so they began to take a serious look at what that would actually mean.

I began my own investigation into the science in 2003, because I found out how much the "solution" would cost and I said that if the United States was even going to consider such expensive, drastic measures that would fundamentally change our economy, the science driving that decision had better be solid. After my rigorous research, I found that it was not—and over the course of six years, more and more flaws continued to surface. Importantly, this was exactly the process that the media and many members of Congress undertook during that year: they realized that we may be seriously considering a solution that would destroy our already ailing economy so they decided to take a harder look at the science—and when they did, they found the man-made catastrophic theory wanting.

By then the Hollywood hysteria had faded into the sunset and the media could no longer indulge in hyped-up fears of climate catastrophe focusing only on the "problem" and ignoring the "solution" as Miles O'Brien did in our 2007 interview when he dismissed my concerns about Kyoto: "We're not talking about Kyoto…We're talking about whether global warming is real." Now that the "solution," was on the horizon, they were forced to face up to it and the economic catastrophe that it would cause. In droves, they started going back to take another look at the science.

In 2007, *Washington Post* staff writer Juliet Eilperin conceded the obvious, writing that climate skeptics "appeared to be expanding rather than shrinking."[226] The mainstream media was finally beginning to take notice. A November 25, 2008, article in *Politico* reported that a "growing accumulation" of science is challenging warming fears, and added that the "science behind global warming may still be too shaky to warrant cap and trade legislation."[227] On October 20, 2008, Canada's *National Post* said that "the number of climate change skeptics is growing rapidly."[228] And *New York Times* environmental reporter, Andrew Revkin, noted on

March 6, 2008, "As we all know, climate science is not a numbers game (there are heaps of signed statements by folks with advanced degrees on all sides of this issue)."[229]

So even before it all collapsed at the end of the year with the Climategate scandal, the science was already well on its way to imploding and the media was starting to catch on to that fact. It was truly the "Year of the Skeptic."

WAXMAN MARKEY PASSES IN THE HOUSE: "WE'LL KILL IT IN THE SENATE"

On Friday, June 26, 2009, Democrats made history. For the first time, a cap and trade bill—sponsored by Representatives Henry Waxman and Ed Markey—passed in the House of Representatives. The House debated the bill all night, and John Boehner stood his ground throwing the 1,400-page document on the desk, saying: "Are we really going to pass a bill that will remake our entire economy in one night, when no one has even read it?"

With the passage of Waxman-Markey that morning, Americans were subjected to yet another significant change in environmental rhetoric. No longer were we moving forward with this effort to avert major climate catastrophe; we were doing this, more importantly, because we needed to transition to a clean energy future. Notably absent in President Obama's ringing endorsement of the bill was any mention of global warming or climate change—or cap and trade for that matter:

> This week, the House of Representatives is moving ahead on historic legislation that will transform the way we produce and use energy in America. This legislation will spark a clean energy transformation that will reduce our dependence on foreign oil and confront the carbon pollution that threatens our planet.
>
> This energy bill will create a set of incentives that will spur the development of new sources of energy, including wind, solar, and geothermal power. It will also spur new energy savings, like efficient windows and other materials that reduce heating costs in the winter and cooling costs in the summer.

These incentives will finally make clean energy the profitable kind of energy. And that will lead to the development of new technologies that lead to new industries that could create millions of new jobs in America — jobs that can't be shipped overseas.

At a time of great fiscal challenges, this legislation is paid for by the polluters who currently emit the dangerous carbon emissions that contaminate the water we drink and pollute the air that we breathe. It also provides assistance to businesses and communities as they make the gradual transition to clean energy technologies.

[...]

We all know why this is so important. The nation that leads in the creation of a clean energy economy will be the nation that leads the 21st century's global economy. That's what this legislation seeks to achieve— it's a bill that will open the door to a better future for this nation. And that's why I urge members of Congress to come together and pass it.[230]

Suddenly we had to pass this bill so that we could reduce carbon "pollution" and create "green jobs"? What happened to saving the world? This kind of language was also the headline of Senator Boxer's first climate change hearing in July 2009 when she said, "Today's hearing is the kickoff of a historic Senate effort to pass legislation that will reduce our dependence on foreign oil, create millions of clean energy jobs, and protect our children from pollution."[231] Just as there was a decided shift in rhetoric from global warming to climate change around 2005–2007, there was a sudden shift in rhetoric from averting climate catastrophe to transitioning to a clean energy economy in 2009. They had to change their strategy to try to counter the reality that cap and trade would destroy jobs.

This did not go unnoticed by the mainstream media. As the *New York Times* put it:

The problem with global warming, some environmentalists believe, is 'global warming.' The term turns people off, fostering images of shaggy-haired liberals, economic sacrifice and complex scientific disputes, according to extensive polling and focus group sessions conducted by ecoAmerica, a nonprofit environmental marketing and messaging firm in Washington.[232]

The *LA Times* also reported:

> Scratch 'cap and trade' and 'global warming,' Democratic pollsters tell
> Obama. They're ineffective...Control the language, politicians know
> and you stand a better chance of controlling the debate. So the Obama
> administration, in its push to enact sweeping energy and healthcare
> policies, has begun refining the phrases it uses in an effort to shape
> public opinion. Words that have been vetted in focus groups and polls
> are seeping into the White House lexicon, while others considered too
> scary or confounding are falling away. [233]

As the *LA Times* rightly points out, cap and trade had also become
a forbidden phrase. Senator Kerry later made his famous statement in
September, "I don't know what 'cap and trade' means. I don't think the
average American does. This is not a cap and trade bill, it's a pollution
reduction bill."[234] I said that if Kerry didn't know what cap and trade
meant, he only had to ask his Democratic colleague in the House, John
Dingell, who had defined it beautifully: "a tax and a big one."[235]

President Obama was clearly moving away from the catastrophe rhetoric, Senator Kerry was pushing green jobs and energy security, and their
ranks were in disarray with the messaging. Several Democrats still hadn't
let go of the idea of preventing bad weather by acts of Congress. Even in
the same opening statement where Senator Boxer headlined with green jobs,
she mentioned toward the end the "devastating effects that will come in the
future if we do not take action to cut global warming pollution. Droughts,
floods, fires, loss of species, damage to agriculture, worsening air pollution
and more." Senator Debbie Stabenow also hadn't let go of the catastrophe,
saying in August 2009, "Climate change is very real. Global warming creates volatility. I feel it when I'm flying. The storms are more volatile. We
are paying the price in more hurricanes and tornadoes."[236]

The final tally of Waxman-Markey's passage in the House portended
its demise. The vote was 219–212, which meant that a large number of
Democrats had voted against it. If a 1,400-page bill that's designed to
transform our economy only passed by seven votes, I knew it didn't stand
a chance in the Senate.

The day before the vote, I was traveling back to Tulsa and stopped by
the *Countrywide and Sun* newspaper in Shawnee, Oklahoma. A reporter

asked me what would happen if the House passed cap and trade and I said, "It doesn't matter because we'll kill it in the Senate anyway."[237]

CAP AND TRADE COMES TO THE SENATE

It was like Lieberman-Warner all over again, except this time Senator Boxer couldn't even get cap and trade to the Senate floor.

From the moment Waxman-Markey passed in the House, the Environment and Public Works Committee under the Chairmanship of Boxer held hearing after hearing on such topics as "Update on the Latest Global Warming Science," "Moving America toward a Clean Energy Economy and Reducing Global Warming Pollution: Legislative Tools," "Clean Energy Jobs, Climate-Related Policies and Economic Growth—State and Local Views," "Climate Change and National Security"—and as before, what Senator Boxer was not as anxious to discuss was cap and trade legislation itself.

But we weren't going to let her get away with that. It didn't matter what the topic of the hearing was. My Republican colleagues and I made it a priority to shine the spotlight on the bill itself and the economic burden it would place on our country. If our friends in the House had debated the bill under the cover of night, we were determined to expose it to the light of day.

At the end of each hearing, I asked my staff to put together a series of YouTube videos that exposed particular faults in cap and trade legislation. There were three pivotal videos that made a significant impact on the debate.

HEARING HIGHLIGHTS: LISA JACKSON CONFIRMS CAP AND TRADE ALL COST, NO GAIN

When President Obama came into office, he brought with him a substantial green team, but I was very pleased that his choice for EPA Administrator, Lisa Jackson, was someone I really liked and respected. Before she joined the Obama Administration, Lisa had testified before the Environment and Public Works Committee on a range of issues from chemical security to mercury legislation when she served as the Director of the New Jersey Environmental Protection Agency office. I had the chance to meet with her again before her nomination hearing, and we talked at length about our families, values, and about the importance of having a healthy respect for our differences of opinion. Lisa left with two items in her hand: my report documenting hundreds of scientists who dispute global warming alarmism and a Christmas card with a picture of my family. She told me that she has that picture hanging on her wall in her office. Lisa and I rarely agree when it comes to environmental policy, but we are good friends and I am always happy to welcome her to the Environment and Public Works Committee.

I especially appreciated her testimony during a particular hearing in July 2009 when she conceded what I had been saying all along: climate change legislation in the United States would be all pain for no gain. At one point during the hearing, I put up a chart of an EPA analysis, which showed that the United States acting alone to reduce greenhouse gas emissions would have no affect on global temperatures, and asked her to confirm that this chart was accurate. She said, "I believe the central parts of the [EPA] chart are that U.S. action alone will not impact world CO_2 levels."[238] One of the reasons I respect Lisa so much is that when you ask her a question, she gives you an honest answer—as she did that day. To have the head of the EPA come out and admit that this bill would have no impact on the climate was a game changer.

Interestingly, just a few weeks before, President Obama in a June 25, 2009, interview covered by the *San Francisco Chronicle*, said, "A long-term benefit [of cap and trade] is we're leaving a planet to our children that isn't four or five degrees hotter." He clearly failed to consult his own EPA on that matter.[239]

HEARING HIGHLIGHTS: GOVERNORS REJECT WAXMAN MARKEY

That July, we also welcomed several governors to testify before the committee for a hearing titled, "Clean Energy Jobs, Climate-Related Policies, and Economic Growth—State and Local Views." Of course, cap and trade was not the intended topic of the hearing, but I made it a point to ask each Governor present how they felt about the Waxman-Markey bill, which put them all in a very uncomfortable position. I knew that Democratic Governor Bill Ritter of Colorado had previously endorsed cap and trade so I was anxious to see if he could still support Waxman-Markey as the governor of a strong oil, gas and agriculture state. Interestingly, when pressed, he could not say that he supported it:

> Here's what I support. I support a national energy policy that's married to a national climate policy that gets at these goals that we have for greenhouse gas reductions. And I believe that if you do that, that there will some vehicle that may not look exactly like Waxman-Markey, particularly after the Senate finishes its work. But I very much support climate legislation that is joined with a national energy policy to get us to the greenhouse gas emission reduction goals that are set for 2050.[240]

Also telling was that Governors Gregoire (D-WA) and Corzine (D-NJ) would not directly say that they endorsed Waxman-Markey.

HEARING HIGHLIGHTS: GREEN JOBS DEBUNKED

During one particularly lively hearing on the clean energy economy also in July, we invited Harry Alford, President and CEO of the National Black Chamber of Commerce, to discuss a new study conducted by the Black Chamber of Commerce and CRA International on the economic effects of Waxman-Markey. As Alford explained:

> Climate change is a vital issue that must be addressed [...] Regretfully, the current legislation out of the U.S. House of Representatives will negatively impact the most vulnerable of our society. I'm sure that those who proposed it had the best intentions, but the bill doesn't do what it's supposed to do, and it does so at a very high cost—especially high for working families and small business owners.[241]

Alford also said that according to this study, "green jobs gained would be swamped by jobs lost in old industries and businesses, leading to a net loss of 2.3 million to 2.7 million jobs," because so many industries supported by fossil fuel development would be destroyed.

Then Senator Boxer let the cat out of the bag. She explained that the goal of her forthcoming legislation was "softening the blow on our trade sensitive industries and our consumers. I just want you to know that, that's the goal."[242] I was glad that she finally admitted that cap and trade would impose significant economic harm. Basically what she was saying was that Washington would tax hard-working Americans, put them out of work, and then cut them a check to help "soften the blow." That didn't sound like a very effective "jobs bill."[243]

THE MOMENT ARRIVES: KERRY-BOXER?

After a long summer of anticipation, Senators Kerry and Boxer finally held an outdoor press conference on September 30, 2009, to introduce their cap and trade bill in the Senate. Right away they faced several problems. First of all, what they introduced that day was an outline—it was not a full bill. Second, it had been clear for a while that a number of conservative Democrats and moderate Republicans in the Senate would not support a bill that was exactly the same as Waxman-Markey, so everyone was wondering what changes they would make on the Senate bill to gain the support of wavering members. Yet, when the bill surfaced, it appeared to be nearly identical to Waxman-Markey. It was also curious that the bill was named Kerry-Boxer rather than Boxer-Kerry since Boxer was, after all, the Chairman of the Environment and Public Works Committee.

CRS REPORT: AMERICA HAS THE LARGEST RESERVES OF FOSSIL FUELS OF ANY COUNTRY IN THE WORLD

On October 23, 2009, just weeks after cap and trade—a bill that would force Americans into policies of energy austerity—came to the Senate, a report from the nonpartisan Congressional Research Service (CRS) revealed for the first time that America's combined energy resources are the largest on earth.[244] Democrats trying to force us off of fossil fuels were not too pleased.

The report tells us that America's combined recoverable natural gas, oil, and coal endowment is the largest on Earth—far larger than those of Saudi Arabia (third), China (fourth), and Canada (sixth) combined. I said, thanks to the non-partisan Congressional Research Service, we now know what resources we have and what we aren't getting out of the ground. I requested the report along with Senator Lisa Murkowski because we had grown tired of the Democrats' refrain that America only has 3 percent of global oil reserves—which, according to this view, meant more drilling and production at home would be futile. The 3 percent mantra is their bread and butter talking point.

President Obama himself has said that with 3 percent of the world's oil reserves, the U.S. cannot drill its way to energy security. My Democratic colleagues in the Senate have also said that the United States has only 3 percent of known oil reserves, yet we use 25 percent of the world's oil production. But the non-partisan CRS shows the full, complete, accurate picture of America's resources—and shows that, yes, we can produce our way to energy security.

That's because CRS shows more than just our proven oil reserves, which is what the Democrats conveniently cite. America's proven reserves, of course, are a modest 28 billion barrels. The word "proven" is important here. The only way to estimate proven reserves is to drill. But that's not possible because federal policies, supported by President Obama and many Democrats, have made of America's federal land inaccessible to drilling.

Knowing what vast resources literally lay at our feet, it seemed all the more outrageous that we were even considering such drastic measures to limit access to these resources.

THE "NUCLEAR OPTION"

Although Kerry-Boxer was "introduced" at the end of September 2009, it took Senators Kerry and Boxer almost a month to draft the full bill, which we received just before midnight on October 23. Senator Boxer planned to begin marking up the bill November 3, 2009, in the Environment and Public Works Committee.

As was the case with Lieberman-Warner, Republicans understandably wanted to have an economic analysis of Kerry-Boxer completed before voting on the bill in committee. Leading the charge on this request was Senator Voinovich from Ohio, a state that would be hit hard by the legislation.

At that stage, EPA had already completed an analysis of Waxman-Markey but they had used unrealistic assumptions—one of them being that the United States would construct 103 new nuclear power plants to make up for lost baseload power from the gradual phasing out of fossil fuels. The idea that we could license and construct that many nuclear power plants was absurd considering that no new power plants have been licensed for thirty years. Senator Voinovich requested that EPA conduct an economic analysis of Kerry-Boxer using more realistic assumptions. But Senator Boxer claimed that our request was only a delay tactic. She said we had enough economic information in our hands to move the bill through Committee.

This disagreement over economic modeling set the stage for a confrontation in the Environment and Public Works Committee that was unprecedented. We stood our ground that we could not move forward with a markup until we had adequate economic modeling of the bill. Senator Boxer decided to hold the markup anyway, breaking long-standing precedent that two members of the minority must be present for a markup to begin. Again, as with Lieberman-Warner, the Copenhagen conference was just around the corner, so Senator Boxer was determined to have this bill passed out of committee to show world leaders that we were taking action.

But as Darren Samuelsohn of *Politico* correctly noted, the consequences for breaking precedent were high: "Going this route [...] could spell trouble for the overall legislation as Boxer and her allies continue their search for 60 votes among moderate Democrats and Republicans. 'That product is totally toxic,' the former staffer warned. 'It's basically worthless.'"[245]

Senator Voinovich attended the markup only on the first day to ask Senator Boxer one last time to allow us to complete the economic analysis before proceeding; when she refused, he left. I briefly attended the last day of the markup to reinforce our message, but she would not concede.

Cap and trade was already dead, but Senator Boxer's decision to break with committee precedent put the last nail in the coffin. I walked straight out of the committee room to the Russell rotunda to give an interview with Martha MacCallum of Fox News on November 5, 2009. I explained how it would all turn out:

INHOFE: We're not going to pass this huge bill. This would be the largest tax increase in the history of America—[we said we would] not pass it out of committee until we had an EPA analysis. We've requested it now for two months. And they're waiting to do it but they won't give them the go ahead. I can only conclude that they don't want the public to know how much money this thing's going to cost. So they've done something that's never been done in the history of this committee. They passed a bill out without one member of the minority—not one Republican. You know, Martha, let's keep in mind all the town meetings that we went through. That was about health care but that was also about this global warming monstrous tax. And I'll tell you this is unprecedented and I think the bill is dead. By the way, one Democrat voted against it and that was Max Baucus. He's from Montana but he's also the chairman of the Finance Committee. So that bill is dead. [...] I really honestly believe this is history in the making right now because it's never happened before. I can't see any way in the world that Harry Reid is going to bring this bill up and it could pass. It just can't happen.[246]

CAP AND TRADE IS DEAD

It wasn't long before I could declare victory. By the middle of November, the Environment and Public Works Committee was back to work on other issues in the hearing room, and I had the chance to have a little fun after Barbara's "elections have consequences" jab. It turns out that votes have consequences too. I was happy to declare, "We won, you lost, get a life!"[247]

6

CLIMATEGATE = VINDICATION

THE COLLAPSE OF THE SCIENCE

When I said I've been called every name in the book, I wasn't kidding. In stark contrast to Rachel Maddow's depiction of me as a "mountain of indignation," Dana Milbank wrote in an October 28, 2009, in *The Washington Post* column, "It must be very lonely being the last flat-earther." It was probably the last time a reporter could get away with singling me out as the only one who wasn't buying into the flawed science behind the global warming campaign. Milbank continued making his case:

"Eleven academies in industrialized countries say that climate change is real; humans have caused most of the recent warming," admitted Sen. Lamar Alexander (R-Tenn.). "If fire chiefs of the same reputation told me my house was about to burn down, I'd buy some fire insurance." An oil-state senator, David Vitter (R-La), said that he, too, wants to "get us beyond high-carbon fuels" and "focus on conservation, nuclear, natural gas and new technologies like electric cars." And an industrial-state senator, George Voinovich (R-Ohio), acknowledged that climate change "is a serious and complex issue that deserves our full attention." Then there was poor Inhofe.[248]

Just a few weeks after that column appeared, it was all over: Climategate, the greatest scientific scandal of our time, broke. So I said Milbank didn't have to feel too sorry for me. What I had been saying about the IPCC all along was confirmed. I was vindicated.

On November 18, 2009, just two days before Climategate, I went back down to the Senate floor to speak about how the "consensus" was already shattered and Copenhagen would fail. I said that 2009 would go down in history as the "Year of the Skeptic." I had a few allies in this assertion: the *Telegraph*, a UK Newspaper, was predicting Copenhagen would be a disaster on November 15, 2009: "The worst kept secret in the world is finally out—the climate change summit in Copenhagen is going to be little more than a photo opportunity for world leaders."[249] I said I would be there to tell them the truth:

And I will be travelling to Copenhagen, leading what I call the "Truth Squad" to say exactly what I said six years ago in Milan, Italy: The United States will not support a global warming treaty that will significantly damage the American economy, cost American jobs, and impose the largest tax increase in American history. Further, as I stated in 2003, unless developing nations are part of the binding agreement, the U.S. will not go along. Given the unemployment rate of 10 percent, and given all of the out of control spending in Washington, the last thing we need is another thousand-page bill that increases costs and ships jobs overseas, all with no impact on climate change.

I also said in Milan that the science is not settled. That was an unpopular view back then. But today, since Al Gore's science fiction movie, more and more scientists, reporters, and politicians are questioning global warming alarmism. I proudly declare 2009 as the "Year of the Skeptic"—the year in which scientists who question the so-called global warming consensus are being heard.[250]

So Copenhagen was already well on its way to failure. When Climategate hit, it only added superfluous nails to a coffin that was already tightly nailed shut.

Climategate revealed leaked emails from the world's top climate scientists at the University East Anglia's Climactic Research Unit, many of whom had been lead authors of the IPCC reports and were intimately involved in writing and editing the IPCC's science assessments. My Senate report showed that many of these scientists may be obstructing the release of information that was contrary to their "consensus" claims; may be manipulating data using flawed climate models to reach preconceived conclusions; may be pressuring journal editors not to publish work questioning the "consensus"; and assuming activist roles to influence the political process.[251]

The implications of this were huge considering that the "consensus" claim was based on the foundation of the IPCC science. Noted science historian Naomi Oreskes wrote, the "scientific consensus" of climate change "is clearly expressed in the reports of the Intergovernmental Panel on Climate Change."[252] One top Obama Administration official said that the IPCC's assessments were the "gold standard" on climate science "because of the rigorous way in which they are prepared, reviewed, and approved."[253]

Each of the IPCC's four assessment reports made the scientific case—more definitely over time—that anthropogenic gases were causing global warming. The IPCC's Fourth Assessment Report's *Summary for Policymakers* in 2007 claimed that "warming of the climate system is unequivocal" and that "[m]ost of the observed increase in globally averaged temperatures since the mid-20th century is very likely due to the observed increase in anthropogenic (human) greenhouse gas concentrations."[254]

Climategate finally destroyed what was left of the façade of the "consensus." Contrary to their repeated public assertions that the "science is settled," the emails show climate scientists were arguing over critical issues, questioning key methods and statistical techniques, expressing concerns about historical periods (such as whether the Medieval Warming Period [MWP] was global in extent) and doubting whether there is "consensus" on the causes and the extent of climate change.

The press reaction in the wake of the scandal was remarkable considering how just a few years before they had nothing but praise for the IPCC. George Monbiot, a British writer, known for his environmental and political activism, wrote in his weekly column in the *Guardian:* "Pretending that this isn't a real crisis isn't going to make it go away. Nor is an attempt to justify the emails with technicalities. We'll be able to get past this only by grasping reality, apologising where appropriate, and demonstrating that it cannot happen again."[255] The *Daily Telegraph* said that "this scandal could well be the greatest in modern science."[256] Clive Crook of the *Atlantic* magazine wrote, "The closed-mindedness of these supposed men of science, their willingness to go to any lengths to defend a preconceived message, is surprising even to me. The stink of intellectual corruption is overpowering."[257]

But comedian Jon Stewart was the best—he said, "Poor Al Gore: global warming completely debunked via the very Internet you invented. Oh the irony!" He went on:

> **STEWART:** Value added data? What is that, numbers fortified with art? Truth plus, now with lemon? It doesn't look good. Now does it disprove global warming? No, of course not. But it does put a fresh set of energizers in the Senate's resident denier bunny.
>
> **SENATOR JAMES INHOFE, (R-OKLA.):** The fact that this whole idea on the global warming. I'm glad that's over, gone, done. We won. You lost. Get a life.
>
> **STEWART:** Alright. We knew Inhofe was going to say that. That guy thinks global warming is debunked every time he drinks a Slushee and gets a brain freeze. "If global warming is real, why does my head hurt?" But by the way, that quote was from BEFORE he found out about the leaked email story. But that's the point. If you care about an issue, and want it to be your life's work, don't cut corners.[258]

It was one of the first times someone called me out for being a "denier" while also giving me credit for predicting how it would all end.

REWIND TO 2005

In 2005, I stood on the Senate floor to discuss the flaws in the IPCC process that had been manifesting themselves for years, and said it was time to face up to the "systematic and documented abuse of the scientific process by which an international body that claims it provides the most complete and objective science assessment in the world on the subject of climate change, the United Nations IPCC."[259]

At the time of my speech, the IPCC's fourth assessment, which was meant to be the "smoking gun" report—attempting to prove there was an "unequivocal" link between humans and catastrophic global warming—was set to come out in 2007.[260] I said that if the IPCC and its fourth assessment were to have any credibility, fundamental changes to the IPCC scientific process would need to be made. Most importantly, I said that the IPCC must adopt procedures that ensure that impartial scientific reviewers formally approve both the chapters *and* the *Summary for Policymakers*—the latter of which was the only document that members of the press and members of Congress ever read. When compared with the *actual* report, it was clear the *Summary for Policymakers* was being co-opted by activists with an agenda to shape the conclusions to show that man-made emissions were causing catastrophic global warming. To safeguard against the manipulation of the message, objective scientists, *not government delegates* should be a part of the approval process. I also said that the IPCC must ensure that any uncertainties in the state of knowledge be clearly expressed in the *Summary for Policymakers*.

But of course, the IPCC remained committed to its path and, as Climategate eventually revealed, it was unsustainable and it was only a matter of time before it collapsed.

THE "GOLD STANDARD"

Phil Jones, the head of the Climatic Research Unit, put it mildly when he admitted that the Climategate emails "do not read well."[261]

The emails themselves raised the important question: what, if any, are the boundaries between science and activism? Perhaps the statement that best exemplifies the unusual political tendency among the scientists in the CRU controversy came from Dr. Keith Briffa, the Deputy Director of the CRU, and lead author of the IPCC's Fourth Assessment Report, who wrote in one of the CRU emails, "I tried hard to balance the needs of the science and the IPCC, which were not always the same."[262] The most famous example comes from an email from Phil Jones, which reads, "I've just completed Mike's Nature trick of adding in the real temps to each series for the last 20 years (i.e. from 1980 onwards) and from 1961 for Keith's to hide the decline."[263] Of course, he means hide the decline in temperatures, which caused another scientist, Kevin Trenberth, to write: "The fact is we can't account for the lack of warming, and it's a travesty that we can't."[264]

Climategate is significant in that it confirmed in the minds of many what we strongly suspected all the time. It is imperative that you read Appendix C, excerpts from our report on the CRU controversy which was published in February 2010. It clearly documents the specific participants and statements in Climategate.

HOCKEY STICK ANNIHILATED

If the hockey stick was already broken all the way back in 2003, after the Climategate revelations, in 2009, it is more accurate to say that it was shattered, as revealed in the following excerpts of my staff report on Climategate emails.

Possibly the most egregious example of scientists trying to silence skeptic voices was the reaction to a paper published in the journal *Climate Research* in 2003, which posed a serious challenge to the "hockey stick" graph constructed by Professors Michael Mann, Raymond Bradley, and Malcolm Hughes. Of course, the hockey stick, which was featured prominently in the IPCC's Third Assessment Report in 2001, supported the conclusion that the 1990s, and 1998, were likely the warmest decade, and the warmest year, respectively, in at least a millennium.

Dr. Sallie Baliunas and Dr. Willie Soon, researchers at the Harvard-Smithsonian Center for Astrophysics, contested the hockey stick conclusion; they reviewed more than two hundred climate studies and "determined that the 20th century is neither the warmest century nor the century with the most extreme weather of the past 1000 years." Their study "confirmed that the Medieval Warm Period of 800 to 1300 A.D.

and the Little Ice Age of 1300 to 1900 A.D., were worldwide phenomena not limited to the European and North American continents. While 20th century temperatures are much higher than in the Little Ice Age period, many parts of the world show the medieval warmth to be greater than that of the 20th century."[265]

As the leaked emails show, Michael Mann, the author of the hockey stick, and Phil Jones, a climatologist at the University of East Anglia, were not too happy about this. In an email on March 11, 2003, titled "Soon and Baliunas," Jones writes that he and his colleagues "should do something" about the Soon-Baliunas study, the quality of which he found "appalling": "I think the skeptics will use this paper to their own ends and it will set paleo [climatology] back a number of years if it goes unchallenged." Jones then went a step further, threatening to shun *Climate Research* until "they rid themselves of this troublesome editor."[266]

That same day, Mann responded, complaining that the skeptics had "staged a bit of a coup" at *Climate Research*, implying that scientists who disagree with him could never get published in peer-reviewed literature solely on the merits of their work. Mann echoed Jones' suggestion to punish *Climate Research* by encouraging "our colleagues in the climate research community to no longer submit to, or cite papers in, this journal:"

> This was the danger of always criticising the skeptics for not publishing in the "peer-reviewed literature." Obviously, they found a solution to that—take over a journal! So what do we do about this? I think we have to stop considering "Climate Research" as a legitimate peer-reviewed journal. Perhaps we should encourage our colleagues in the climate research community to no longer submit to, or cite papers in, this journal. We would also need to consider what we tell or request of our more reasonable colleagues who currently sit on the editorial board.[267]

In April 2003, Timothy Carter with the Finnish Environment Institute suggested changes to the editorial process at *Climate Research* in an email to Tom Wigley, a scientist formerly with the University Corporation for Atmospheric Research (UCAR). Noting communications with "Mike" (Michael Mann) the previous morning, Carter wondered how to remove "suspect editors," presumably those who approve research by skeptics. In reply, Wigley described a campaign to discredit *Climate Research* through a letter signed by more than fifty scientists. He also mentioned Mann's approach to "get editorial board members to resign":

> One approach is to go direct to the publishers and point out the fact that their journal is perceived as being a medium for disseminating misinformation under the guise of refereed work. I use the word "perceived" here, since whether it is true or not is not what the publishers care about—it is how the journal is seen by the community that counts. I think we could get a large group of highly credentialed scientists to sign such a letter—50+ people. Note that I am copying this view only to Mike Hulme and Phil Jones. Mike's idea to get editorial board members to resign will probably not work—must get rid of von Storch too, otherwise holes will eventually fill up with people like Legates, Balling, Lindzen, Michaels, Singer, etc. I have heard that the publishers are not happy with von Storch, so the above approach might remove that hurdle too.[268]

Along with these discussions about removing journal editors who held contrary views on climate science, the emails show that the scientists tried to prevent publication of papers they disagreed with. On July 8, 2004, Jones suggested that he and a colleague could keep the work of skeptics from appearing in the IPCC's Fourth Assessment report: "I can't see either of these papers being in the next IPCC report. Kevin and I will keep them out somehow—even if we have to redefine what the peer-review literature is!"[269]

The conclusion is obvious: Mann and his colleagues were not disinterested scientists. They acted more like a priestly caste, viewing substantive challenges to their work as heresy. And rather than welcoming criticism and debate as essential to scientific progress, they launched a campaign of petty invective against scientists who dared to question their findings and methods.

"HIDE THE DECLINE"

The following is more from my Environment and Public Works Committee's minority staff report on Climategate:

> "I am not sure that this unusual warming is so clear in the summer responsive data. I believe that the recent warmth was probably matched

about 1,000 years ago." Keith Briffa, Deputy Director, CRU, September 22, 1999.

I asked what Dr. Mann was trying to hide in a speech on the Senate floor on April 13, 2005, and Climategate emails provided the answer: he was arguably hiding the decline in temperatures. One of the most famous emails is written by CRU's Jones in 1999: "I've just completed Mike [Mann]'s Nature trick of adding in the real temps to each series for the last 20 years (i.e. from 1981 onwards) and from 1961 for Keith's to hide the decline."[270]

Jones's "trick" arose because of disagreement over Dr. Mann's "hockey stick" temperature graph. Of course, the hockey stick showed a relatively straight shaft extending from 1000 AD to 1900, when a blade turns sharply upward, suggesting that warming in the 20th century was unprecedented, and caused by anthropogenic sources. Remember, the hockey stick was featured prominently on page one of the IPCC's *Summary for Policymakers* in its Third Assessment Report.

In defending himself, Jones said, "The word 'trick' was used here colloquially as in a clever thing to do. It is ludicrous to suggest that it refers to anything untoward."[271] Similarly, echoing Jones, Dr. John Holdren, President Obama's Science Adviser, asserted that "trick" merely means a clever way to tackle a problem.[272] Both Holdren and Jones' explanation of "trick" used in this context has evidentiary support. Unfortunately, neither Jones nor Holdren addressed the "problem" that confronted Jones and his colleagues. The problem in this case is the so called "divergence problem." The divergence problem is the fact that after 1960, tree ring reconstructions show a marked decline in temperatures, while the land-based, instrumental temperature record shows just the opposite.

For some scientists, the divergence of data was a cause of great concern, but not necessarily for scientific reasons. For instance, IPCC author Chris Folland warned in an email that such evidence "dilutes the message rather significantly" that warming in the late 20th century relative to the last 1,000 years is "unprecedented."[273]

Specifically, Jones et al. expressed concern about a temperature reconstruction authored by Keith Briffa, a senior researcher with CRU. Because reliable thermometer data go back only to the 1850s, scientists use proxy data such as tree rings to reconstruct annual temperatures over long periods (e.g., 1,000 years) (it must be noted that proxy reconstructions are rife with uncertainties).[274]

Unfortunately for those in the email chain, Briffa's reconstruction relied on tree ring proxies that produced a sharp and steady decline in temperature after 1960. This conflicted with the instrumental temperature readings that showed a steep rise. Briffa's graph was, according to Dr. Michael Mann, a "problem":

> Keith's series…differs in large part in exactly the opposite direction that Phil's does from ours. This is the problem we all picked up on (everyone in the room at IPCC was in agreement that this was a problem and a potential distraction/detraction from the reasonably consensus viewpoint we'd like to show w/ the Jones et al and Mann et al series.[275]

Briffa later addressed the "pressure to present a nice tidy story" about the "unprecedented" warming in the late 20th century. In his view, "the recent warmth was matched about 1,000 years ago." Here is the email from Briffa in full:

> I know there is pressure to present a nice tidy story as regards 'apparent unprecedented warming in a thousand years or more in the proxy data but in reality the situation is not quite so simple. We don't have a lot of proxies that come right up to date and those that do (at least a significant number of tree proxies) some unexpected changes in response that do not match the recent warming. I do not think it wise that this issue be ignored in the chapter. For the record, I do believe that the proxy data do show unusually warm conditions in recent decades. I am not sure that this unusual warming is so clear in the summer responsive data. I believe that the recent warmth was probably matched about 1,000 years ago. I do not believe that global mean annual temperatures have simply cooled progressively over thousands of years as Mike appears to and I contend that there is strong evidence for major changes in climate over the Holocene (not Milankovich) that require explanation and that could represent part of the current or future background variability of our climate.[276]

Mann was apparently nervous that "skeptics" would have a "field day" if Briffa's decline was featured in the IPCC's Third Assessment Report. He said "he'd hate to be the one" to give them "fodder."

On September 22, 1999, Mann wrote:

> We would need to put in a few words in this regard. Otherwise, the skeptics have a field day casting doubt on our ability to understand the factors that influence these estimates and, thus, can undermine faith in the paleoestimates. The best approach here is for us to circulate a paper addressing all the above points. I'll do this as soon as possible. I don't think that doubt is scientifically justified, and I'd hate to be the one to have to give it fodder![277]

As UK's *Daily Mail* reported, "All [Jones] had to do was cut off Briffa's inconvenient data at the point where the decline started, in 1961, and replace it with actual temperature readings, which showed an increase."[278]

So it seems that, rather than employing a "clever way"—or "trick"—to solve the post-1960 decline, Jones allegedly manipulated data to reach a preconceived conclusion. His method has been criticized by fellow scientists. Philip Stott, emeritus professor of biogeography at London's School of Oriental and African Studies said, "Any scientist ought to know that you just can't mix and match proxy and actual data. They're apples and oranges. Yet that's exactly what [Jones] did."[279]

"RECKLESS ENDANGERMENT"

What they couldn't achieve through Kyoto they tried to achieve through cap and trade legislation. And what they couldn't achieve through legislation, they are currently trying to achieve through regulation.

In the midst of the cap and trade debate, as support for the bill was dwindling, the Obama EPA was working behind the scenes to finalize a "finding" that greenhouse gases harm public health and welfare, known as the "endangerment finding." During a key case, *Massachusetts v. EPA*,[280] the Supreme Court ruled that *if* EPA determined that greenhouse gases endanger human health, *then* they must regulate them under the Clean Air Act. The key word here is "if." Proponents of the endangerment finding claim that the court forced the EPA to move forward with this finding, but this is not the case. The courts were clear that the EPA Administrator first had to determine *if* greenhouse gases endanger the public, and that determination would require a scientific investigation. They had a choice, and they made the wrong choice. They chose to make an endangerment finding based on the flawed scientific conclusions of the IPCC.

As the cap and trade battle waged on, EPA Administrator Lisa Jackson and President Obama said repeatedly that passing the bill was preferable because EPA regulations would be much more costly and complicated. As one astute April 2009 editorial in the *Wall Street Journal* put it, the Administration essentially played Russian roulette with regulations:

> President Obama's global warming agenda has been losing support in Congress, but why let an irritant like democratic consent interfere with saving the world? So last Friday the Environmental Protection Agency decided to put a gun to the head of Congress and play cap and trade roulette with the U.S. economy.
>
> The pistol comes in the form of a ruling that carbon dioxide is a dangerous pollutant that threatens the public and therefore must be regulated under the 1970 Clean Air Act. This so-called 'endangerment finding' sets the clock ticking on a vast array of taxes and regulation that EPA will have the power to impose across the economy, and all with little or no political debate.[281]

They were determined to have cap and trade no matter what.

Part of the reason EPA regulation of greenhouse gases would be more complicated is that the Clean Air Act thresholds are only meant to regulate real, localized pollutants such as SO_2, NOX, and Mercury. Numerous legal experts, including Democrat Representative John Dingell, who wrote the Clean Air Act amendments, have said that the Clean Air Act was never designed to regulate greenhouse gas emissions. That's because emissions of greenhouse gases are far greater than conventional pollutants, if EPA regulated them at the thresholds required by the Clean Air Act, the Agency would have to regulate almost everything including schools, hospitals, nursing homes, commercial buildings, churches, restaurants, hotels, malls, colleges and universities, food processing facilities, farms, sports arenas, soda manufacturers, bakers, brewers, wineries, and even some private homes. The results of that would be absurd, so EPA tailored the Clean Air Act to create much higher thresholds for greenhouse gases—but this tailoring will not likely hold up in the courts because it directly contradicts the law. And if the courts throw the "tailoring rule" out, it will be as Representative John Dingell put it, "a glorious mess."[282]

And the entire foundation of this bureaucratic nightmare is flawed science: EPA Administrator Lisa Jackson admitted to me publicly that EPA based its action on the IPCC science, saying that the proposal, the Agency relied in large part on the assessment reports developed by the Intergovernmental Panel on Climate Change (IPCC).

At an Environment and Public Works Committee hearing on December 2, 2009, I challenged Administrator Jackson on that matter, saying that given what has come to light in the Climategate scandal, EPA should halt this agenda. She replied, "While I would absolutely agree that these emails show a lack of interpersonal skills, as I would say to my kids, be careful who you write, and maybe more, I have not heard anything that causes me to believe that that overwhelming consensus that climate change is happening and that man-made emissions are contributing to it, have changed."[283] At an Environment and Public Works Committee hearing in February 2010, Administrator Jackson told me that EPA accepted the findings of the IPCC without any serious, independent analysis.[284]

Directly in line with Administrator Jackson's points, the endangerment finding states, "it is EPA's view that the scientific assessments" of the IPCC "represent the best reference materials for determining the general state of knowledge of the scientific and technical issues before the agency in making an endangerment decision."[285] In the finding's Technical Support Document (TSD), in the section on "attribution," EPA claims that climate changes are the result of anthropogenic greenhouse gas emissions and not natural forces. In this section EPA has 67 citations, 47 of which refer to the IPCC.[286]

If there are any objective readers of this book who still give credibility to the IPCC science on which the entire hoax is based and the basis of the endangerment finding, then reading Appendix C is a must.

DR. ALAN CARLIN

Rewind for a moment to March 9, 2009, when Dr. Alan Carlin, a PhD economist in EPA's Office of Policy, Economics, and Innovation released a key report that called into question the scientific process underlying the agency's proposed endangerment finding. According to Carlin, a thirty-eight-year veteran of EPA, and a fellow agency employee:

> We have become increasingly concerned that EPA and many other agencies and countries have paid too little attention to the science of global warming. EPA and others have tended to accept the findings reached by outside groups, particularly the IPCC and the CCSP, as being correct without a careful and critical examination of their conclusion and documentation...We believe our concerns and reservations are sufficiently important to warrant a serious review of the science by EPA before any attempt is made to reach conclusions on the subject.[287]

But Carlin's request was denied. In a series of emails, Al McGartland, Carlin's boss, forbade him from having "any direct communication" with anyone outside of his office concerning his study. On March 16, Carlin tried again, but McGartland made clear what his superiors thought of the report: "The time for such discussion of fundamental issues has passed for this round. The administrator and the [Obama] administration have decided to move forward on endangerment, and your comments do not help the legal or policy case for this decision... I can only see one impact of your comments given where we are in the process, and that would be a very negative impact on our office." But that wasn't all. McGartland also wrote to Carlin: "With the endangerment finding nearly final, you need to move on to other issues and subjects. I don't want you to spend any additional EPA time on climate change. No papers, no research etc., at least until we see what EPA is going to do with Climate."[288] So much for transparency.

The endangerment finding was finalized December 7, 2009, just in time for the Copenhagen climate conference.[289] Cap and trade had failed, but President Obama could still face world leaders armed with his back-up plan. The cost of "doing something" for the conference was high: the endangerment finding, like cap and trade would cost American consumers around $300 to $400 billion a year, significantly raise energy prices, and destroy hundreds of thousands of jobs.

CRISIS OF CONFIDENCE IN THE IPCC

After Climategate there was an interesting reversal in the mainstream media: all those outlets that had praised Al Gore and the IPCC to the heights just a few years prior were suddenly tearing apart the IPCC's assessments—and more and more and more flaws came to light. When *ABC News*,[290] the *Economist*,[291] *Time*,[292] *Newsweek*,[293] and the *Financial Times*[294]—among many others—reported that the IPCC's research contains embarrassing flaws, and that the IPCC chairman and scientists knew of the flaws, but published them anyway—well, you have the makings of a major scientific scandal. In the end, well over a hundred different errors in the IPCC science were revealed in the wake of the Climategate email scandal.

One of the most publicized errors was, of course, IPCC's claim that the Himalayan glaciers would melt by 2035. It's simply false, yet it was put into the IPCC's Fourth Assessment report. Here's what we know:

- According to the *Telegraph*, "the IPCC [has] since admitted it was based on a report written in a science journal and even the scientist who was the subject of the original story admits it was not based on facts."[295]

- "When finally published," the *Telegraph* wrote, "the IPCC report did give its source as the WWF study but went further, suggesting the likelihood of the glaciers melting was 'very high'." (The IPCC, by the way, defines this as having a probability of greater than 90%.)[296]

Time magazine, the very publication that once told us to be afraid— very afraid of global warming, said that "Glaciergate," was a "black eye for the IPCC and for the climate-science community as a whole."[297]

There was more. According to the *Telegraph*, Dr. Rajendra Pachauri, the head of the IPCC, "was informed that claims about melting Himalayan glaciers were false before the Copenhagen summit.[298]

So why was the Himalayan error included? We now know from the very IPCC scientist who edited the report's section on Asia that it was done for political purposes—it was inserted to induce China, India, and other countries to "take action" on global warming. According to the UK's *Sunday Mail*, Murari Lal, the scientist in charge of the IPCC's chapter on Asia, said "We thought that if we can highlight it, it will impact policymakers and politicians and encourage them to take some concrete action." In other words, as the *Sunday Mail* wrote, Lal "admitted [the glacier alarmism] was included purely to put political pressure on world leaders."[299]

So what had the IPCC done to rectify this fiasco? I went into the IPCC report to see if a correction had been made: the 2035 claim was still there. Of course, there was a note attached and it said the following:

> It has, however, recently come to our attention that a paragraph in the 938-page Working Group II contribution to the underlying assessment refers to poorly substantiated estimates of rate of recession and date for the disappearance of Himalayan glaciers. In drafting the paragraph in question, the clear and well-established standards of evidence, required by the IPCC procedures, were not applied properly.[300]

It turns out that the IPCC's fourth assessment also found observed reductions in mountain ice in the Andes, Alps, and Africa—all caused by, of course, global warming. In an article titled, "UN climate change panel based claims on student dissertation and magazine article," the *Telegraph* reported:

> . . . one of the sources quoted was a feature article published in a popular magazine for climbers which was based on anecdotal evidence from mountaineers about the changes they were witnessing on the mountainsides around them. The other was a dissertation written by a geography student, studying for the equivalent of a master's degree, at the University of Berne in Switzerland that quoted interviews with mountain guides in the Alps.[301]

On top of this, we found that the IPCC was exaggerating claims about the Amazon. The report said that 40 percent of the Amazon rainforest was endangered by global warming. But, again, as we've seen, this was taken from, yes, a study by the World Wildlife Federation, and one that had nothing to do with global warming. Even worse, it was written by a green activist.

In the wake of the scandal, even my good friend Barbara Boxer was careful about how she talked about the IPCC. As she said in a hearing on EPA's Budget on February 23, 2010, "In my opening statement, I didn't quote one international scientist or IPCC report...We are quoting the

American scientific community here."

This was the "gold standard" of climate research; it was the body that was awarded the Nobel Peace Prize in 2007. It obviously did not win a Nobel science award.

ONLY A MATTER OF TIME

This crisis of confidence in the IPCC translates to a crisis of confidence for EPA's endangerment finding, which rests in large measure on the IPCC's conclusions. The endangerment finding's scientific foundation has already disintegrated and I believe it will only be a matter of time before the finding itself follows suit.

Once I had it confirmed from Lisa Jackson that the EPA had relied on the science of the IPCC to establish the endangerment finding, I asked the EPA Office of Inspector General (OIG) in April 2010 to investigate the process leading up to the endangerment finding to determine if EPA had come to that conclusion properly. In September 2011, the OIG completed its report and found that the EPA had not come to this conclusion properly—in fact, it found that the scientific assessment underpinning the Obama EPA's endangerment finding for greenhouse gasses was inadequate and in violation of the Agency's own peer review process.[302] The report calls the scientific integrity of EPA's decision-making process into question and undermines the credibility of the endangerment finding.

The Inspector General's investigation uncovered that the EPA failed to engage in the required record-keeping process leading up to the endangerment finding decision, and it also did not follow its own peer review procedures to ensure that the science behind the decision was sound. Regardless of what one thinks of the UN science, the EPA is still required—by its own procedures—to conduct an independent review. Dr. Alan Carlin is now vindicated as his concerns that EPA was relying too much on the science of the IPCC, and that the Agency was not engaging in a rigorous scientific process, turned out to be valid.

I was reminded of Jon Stewart's warning "don't cut corners!" when the press began weighing in on the IG report. The headline from the AP was "U.S. watchdog: EPA Took Shortcut on Climate Finding."[303] EPA immediately responded saying, as it did during the Climategate scandal, that this still does not affect the validity of the science, but Stewart's admonition of the scientists of Climategate applies to the EPA: if the Agency is so sure the science is completely sound, why did they cut corners? Why can't they be transparent if there is nothing to hide?

EPA's process to determine the endangerment finding was rushed, biased, and appears to have been predetermined. Now that all of this has come to light, the only conclusion is that the endangerment finding should be thrown out, and if it is, it will be a tremendous victory for the American people.

7

COPENHAGEN

Speaking to reporters in Copenhagen

BACK INTO THE LION'S DEN

I was only in Copenhagen for three hours but they were the most exhilarating three hours of my political life.

I landed at 7 a.m. on a particularly frigid morning to be holding a conference on global warming. Because I was coming to Copenhagen as a Senator in the minority party, on my own, and not part of any Congressional Delegation, the UN Climate Conference would not give me a meeting room to visit with UN officials or hold a press conference. But with the President planning to make an appearance, over one hundred members of the press were locked in a press room so I saw the perfect opportunity to deliver my message. There was a staircase at the front of the press room which gave me the best vantage point. Nothing was going on at the time: Al Gore and Senator John Kerry had just left—so even if the press wasn't very happy to see me, at least my visit was some news. I know it sounds strange to say it but the experience was really quite enjoyable. I will always remember all those people in the room—hundreds of them—and all the cameras. And they all had one thing in common: they all hated me.

Fox News had it right when they said I had just walked into the "lion's den":

> Inhofe is the second member of congress to arrive in Copenhagen. Kerry addressed reporters last night and got a standing ovation inside the Bella Convention Center. By contrast, Inhofe was mobbed inside the press center and was not offered a speaking slot...Surrounded by reporters from around the world, including many who believe global warming is real, Inhofe often looked like a lamb on his way to slaughter.[304]

I was just happy to see that I had maintained my distinction of being the "most dangerous man on the planet" after all these years. As *Der Spiegel* would later write:

> He only came for a few hours, but he was also decisively responsible for the chaos that marked the negotiations: James Inhofe, a Republican politician who does whatever he can in Washington to inhibit Obama's efforts to impose CO_2 limits. He is not only ridiculous in describing climate change as made up by "the Hollywood elite," but outright dangerous...Men like Inhofe, who in Copenhagen warned that nations shouldn't be "deceived into thinking the US would pass cap and trade legislation," have the effect of poison when it comes to the urgently needed global trust-building.[305]

I still find it amazing that they thought me capable of single-handedly destroying the planet.

ONE-MAN TRUTH SQUAD

Leading up to the conference, President Obama was trying to make it sound like there was still hope that the United States would act on global warming. On September 22, 2009, he made a point to highlight the efforts of cap and trade legislation in Congress as if that were going to lead to something:

> The House of Representatives passed an energy and climate bill in June that would finally make clean energy the profitable kind of energy for American businesses and dramatically reduce greenhouse gas emissions. One committee has already acted on this bill in the Senate and I look forward to engaging with others as we move forward.[306]

But by November 17, 2009, *Wall Street Journal* said it best in their editorial entitled, "Copenhagen's Collapse—The Climate Change Sequel Is a Bust":

> "Now is the time to confront this challenge once and for all," President-elect Obama said of global warming last November. "Delay is no longer an option." It turns out that delay really is an option-the only one that has world-wide support. Over the weekend Mr. Obama bowed to reality and admitted that little of substance will come of the climate-change summit in Copenhagen next month. For the last year the President has been promising a binding international carbon-regulation treaty a la the Kyoto Protocol, but instead negotiators from 192 countries now hope to reach a preliminary agreement that they'll sign such a treaty when they meet in Mexico City in 2010. No doubt. The environmental lobby is blaming Copenhagen's pre-emptive collapse on the Senate's failure to ram through a cap and trade scheme like the House did in June, arguing that "the world" won't make commitments until the U.S. does. But there will always be one excuse or another, given that developing countries like China and India will never be masochistic enough to subject their economies to the West's climate neuroses. Meanwhile, Europe has proved with Kyoto that the only emissions quotas it will accept are those that don't actually have to be met.[307]

Then as Anna Fifield of the *Financial Times* put it in a story called "U.S. Senator Calls Global Warming a Hoax," on December 5, 2009, even

President Obama had "conceded that the Copenhagen summit is not going to result in a binding international treaty."[308] Yet he was still planning to commit the United States to a 17 percent reduction in greenhouse gas emissions from 2005 levels by 2020, which were the cuts prescribed in the Waxman-Markey bill.

The President was clearly in a difficult position. He had failed even to bring together his own party in the Senate to pass the cap and trade bill, so there was no chance he would bring the world together to implement an international treaty as he had once promised. For weeks there was plenty of speculation as to whether President Obama would even attend in person due to the fact that the effort was going to fail, but just days before the conference we learned that the President would indeed be traveling to Copenhagen to commit the United States to emissions cuts and deliver billions of taxpayer dollars to the global warming effort.

I had the opportunity to debate the President's plan with Ed Markey on *FOX News Sunday* with Chris Wallace the weekend before the Conference:

CHRIS WALLACE, HOST: Senator Inhofe, in Copenhagen, the president is reportedly going to pledge the U.S. will reduce greenhouse gas emissions by 17 percent by the year 2020 and will contribute billions of dollars to developing countries to help them reduce their emissions. How much authority will the president's pledge have?

SEN. JAMES INHOFE, (R-OK): Well, see, Chris, that's the reason I'm going, to make sure people in these other 191 countries know the president can't do that.

The initial reductions he's talking about are what you find in Markey's bill, and that isn't going to happen. And of course, that bill's dead. It will never even be brought up again.

And on top of that, he's going to commit, I understand, to some $10 billion a year to the developing countries. Now, here's China that holds $800 billion of our debt and we're going to give them $10 billion to stop generating electricity? I don't think that's going to happen.

Representative Markey, however, was in denial. As he said:

REP. ED MARKEY, (D-MA): So as the president goes there, based upon the Waxman-Markey bill, which has passed through the House of Representatives, which is a 17 percent reduction of greenhouse gases by 2020—the bill that's already passed through the Senate Environment Committee, which is also a 17 to 20 percent reduction, combined with all of the other activity that Senator Lindsey Graham, Senator Joe Lieberman, Senator John Kerry...

WALLACE: Yeah, but none of this has passed.

MARKEY: Senator Susan Collins are all moving towards, there is real momentum now building for a bipartisan bill to pass through the United States Senate.[309]

Especially after that interview, I felt that I needed to travel to Copenhagen, as the left was in full media spin cycle and clearly in denial. But as I had predicted in a *New York Times* article in November, Senate votes were scheduled on the healthcare bill conveniently at the time of the Copenhagen conference, so that cast doubt on the attendance of any senators. Senator Boxer cancelled her trip and instead gave a video address to the conference. Given the significance of the votes on the healthcare bill and believing my colleagues would not attend, I met with the rest of my Truth Squad and we decided to cancel.

However, I found out later that Senator Kerry was going to sneak over to the conference after all. Once I heard that, I knew I had to be there as well because Senator Kerry would not tell them the truth. I immediately booked a flight that would have me on the ground in Copenhagen for only three hours. I scheduled it so that my message could be delivered and still be back in Washington in time for some key votes.

And as I predicted, Senator Kerry claimed in an interview that I was wrong and cap and trade still had a shot:

REPORTER: Senator Inhofe's staff are here meeting with a number of foreign delegations. He's coming tomorrow. He's giving a lot of interviews to foreign media saying essentially this is going to be a repeat of Kyoto, that we cannot pass this.

SENATOR KERRY: Well he's wrong, he's just dead wrong. He's wrong. We have a president of the United States that isn't George Bush, and we have a Senate that doesn't reflect Senator Inhofe's view. Senator Inhofe does not accept the science.

REPORTER: But how might the rest of the world take that message?

SENATOR KERRY: I think everybody understands there are doubters and skeptics and there are people, as I said, who accept the science. The vast majority of people in the world who understand the science accept it. And they want action. I don't think he represents the majority. What he represents is the fight that we have over these issues but I don't think he represents the majority on this issue.

REPORTER: The latest numbers show that people don't believe—in America—that this is as much a problem as they've been led to believe. And that they don't want to have to change the way they live...those numbers are going south.

SENATOR KERRY: You don't have to change the way you live. Nobody is suggesting...nobody has to give up any comfort. Nobody has to change the way they live. They can live better. They can live safer, live healthier, live with more income in their household, live with less money going to wasteful energy. This is a better road for the creation of jobs and the security of our country and I believe that, again, that most Americans understand that.[310]

Senator Kerry's statement is interesting not only because it turned out that he was the one who was "dead wrong" about cap and trade, but because it also clearly exemplifies the crisis of messaging from which the global warming alarmists continually suffered. It was almost impossible to keep track of all the different ways global warming was sold: first they told us that the planet was warming to a dangerous degree and we were going to be wiping sweat off our brows in the middle of February; then the message was that cold weather and every weather event was due to global warming; then Al Gore asked us in his movie if we were ready to change the way we live, while the Hollywood elite went on an extensive campaign to tell us to take public transportation, avoid planes and cars, and change our light bulbs; then the debate was no longer about saving the world but about transitioning to a clean energy economy, achieving energy security, and reducing pollution; then in one desperate final attempt, Senator Kerry tells us that we don't have to change the way we live or give up any comfort. In fact, according to Senator Kerry, cap and trade will make us a more prosperous nation.

While my environmental friends changed their tune constantly throughout the years to justify imposing the largest tax increase in American history, and as each one of their rhetorical tactics failed with the American people, my message has been exactly the same from the very beginning: the science is not settled; their solutions—whether they be Kyoto, cap and trade, or global warming regulations by the EPA—are only symbolic and will only be all economic pain for no environmental gain; and as I said first in Milan and again in Copenhagen, the United States would never ratify Kyoto or impose cap and trade. On the steps of the press room, I told them the same thing I've been saying all along:

I want to turn back the clock to December 2003, when the United Nations convened the "9th Conference of the Parties" in Milan, Italy, to discuss implementation of the Kyoto Protocol. At the time, I was leading the Senate delegation to Milan as Chairman of the Senate Committee on Environment and Public Works. Fast forward to December 2009: the UN is holding its 15th global warming conference-and the delegates are haggling over the same issues that were before them in 2003. I know this because I was there. Recently, with the Copenhagen talks underway, I reread the speech I delivered in Milan. I found that the issues at stake in 2003 are nearly the same as those in 2009. In short, nothing has changed and nothing has been done.

So let's go back to 2003. In my speech, I told the conference that the Senate would not ratify Kyoto. Here's what I said: "The Senate, by a vote of 95 to 0, approved the Byrd-Hagel resolution, which warned the President against signing a treaty that would either economically harm the United States or exempt developing countries from participating." I went on to say this: "Both those conditions then, and still to this day, have not been satisfied. So, it's worth noting that even if President Bush wanted to submit the treaty to the Senate, it couldn't be ratified." That was 2003.

Is that still true today? Of course it is. And yet here we go again: China, India, and other developing countries want nothing to do with absolute, binding emissions cuts. China and India have pledged to reduce the rate of growth, or intensity, of their emissions. But that's not acceptable to the US Senate. Moreover, China is opposed to a mandatory verification regime to prove it is actually honoring its commitments.

[...]

My stated reason for attending Copenhagen was to make certain the 191 countries attending COP-15 would not be deceived into thinking the US would pass cap and trade legislation. That won't happen. And for the sake of the American people, and the economic well-being of America, that's a good thing.[311]

I may have been the most hated man at the conference, but I was the only one that was willing to tell them the truth. As I joked later on Wolf Blitzer's show, "They're in the middle of kind of a group therapy right now."[312]

LEAVING THEIR FOOTPRINT

In the end, President Obama's "agreement" was one that did not bind the United States to any new targets or timelines, with no way to verify or enforce any type of emission cuts. The entire effort had failed, and that was a victory for the American people.

But as CBS reported, the United States made its mark in other ways:

Few would argue with the U.S. having a presence at the Copenhagen Climate Summit. But wait until you hear what we found about how many in Congress got all-expense paid trips to Denmark on your dime. CBS investigative correspondent Sharyl Attkisson reports that cameras spotted House Speaker Nancy Pelosi at the summit. She called the shots on who got to go. House Majority Leader Steny Hoyer, and embattled Chairman of the Tax Committee Charles Rangel were also there...

Senator Inhofe (R-OK) is one of the few who provided us any detail. He attended the summit on his own for just a few hours, to give an "opposing view." "They're going because it's the biggest party of the year," Sen. Inhofe said. "The worst thing that happened there is they ran out of caviar."

Nobody we asked would defend the super-sized Congressional presence on camera. One Democrat said it showed the world the U.S. is serious about climate change. And all those attendees who went to the summit rather than hooking up by teleconference? They produced enough climate-stunting carbon dioxide to fill 10,000 Olympic swimming pools. Which means even if Congress didn't get a global agreement—they left an indelible footprint all the same.[313]

8

THE ATTEMPTS TO RAISE CAP AND TRADE FROM THE GRAVE

"One of the things I don't think happens often enough in our society in part because it doesn't happen so often that we have public figures who stand up who set their feet squarely forward and say this is nonsense. We have to be fact based, we have to be rational and this nonsense has to end. James Inhofe has been such a man over the past six to seven years. He sometimes stood absolutely alone and was demonized, vilified, ridiculed by the national media. He stands now in 2010 as a man utterly vindicated."

—Lou Dobbs, February 24, 2010[314]

BY THE END OF 2009, cap and trade was dead and buried with so many nails in the coffin, yet Senator Kerry, Senator Graham, and Senator Lieberman still got their shovels out and started trying to dig it out in a fruitless attempt to resurrect the bill.

In November 2009, even as Senator Boxer was marking up in Committee her version of cap and trade, Kerry-Boxer, without any Republicans, Senator Kerry was already in talks about introducing a separate cap and trade bill—he called it a "dual track" measure.[315] He undoubtedly saw the writing on the wall that Kerry-Boxer was failing and he didn't want to go to the UN Copenhagen climate conference without a Plan B. So while Obama faced national leaders with the endangerment finding in hand, Kerry came armed with the promise of a bill that he would unveil with Senators Graham and Lieberman. On December 10, 2009, the three presented not their bill but the framework for a bill[316]—it sounded just like the process for Kerry-Boxer. They were trying to distract the public with multiple bills in a "smoke and mirrors" campaign. Here we go again.

This tripartisan trio's plan was to implement the same targets as the Waxman-Markey bill: a 17 percent reduction by 2020. Their strategy right away was to say that this was the "last call" before regulations would set in. Of course, President Obama and EPA Administrator Lisa Jackson often reminded us that regulations under the Clean Air Act would be much worse than a cap and trade bill. Senators Kerry and Graham echoed this point in a joint *New York Times* op-ed:

> Failure to act comes with another cost. If Congress does not pass legislation dealing with climate change, the administration will use the Environmental Protection Agency to impose new regulations. Imposed regulations are likely to be tougher and they certainly will not include the job protections and investment incentives we are proposing.
>
> The message to those who have stalled for years is clear: killing a Senate bill is not success; indeed, given the threat of agency regulation, those who have been content to make the legislative process grind to a halt would later come running to Congress in a panic to secure the kinds of incentives and investments we can pass today. Industry needs the certainty that comes with Congressional action.[317]

And, as with Kerry-Boxer, they didn't want to call it cap and trade, even though that's exactly what it was. As Steve Benen of the *Washington Monthly* aptly put it:

> Apparently, "cap and trade" no longer polls well. The White House seems to now prefer "energy independence legislation." These three senators are using the phrase "market-based approach." ("You remember the artist formerly known as Prince?" Lieberman said. "This is the market-based system for punishing polluters previously known as 'cap and trade.'")[318]

More "smoke and mirrors." Months later ,Senator Reid reinforced this statement saying that even though the bill does create caps on greenhouse gases, he doesn't like the term "cap and trade." As he said, "We don't use the word 'cap and trade.' That's something that's been deleted from my dictionary. Carbon pricing is the right term."[319]

While Kerry-Boxer was branded as a "pollution reduction bill," Kerry-Graham-Lieberman was going to be "energy independence legislation" that will "punish polluters"—only the polluters would not be punished; even by President Obama's own admission, it would be the American people who would be punished. When the President said himself that under his plan of a cap and trade system "electricity rates would necessarily skyrocket" he explained that it was because power plants "would have to retro-fit their operations. That will cost money. They will pass that money on to the consumers."[320] No wonder they didn't like to say cap and trade.

In fact, not only would the "polluters" not be punished, they were cutting some pretty good deals for themselves.

THE APPEASERS

When I think of all the backroom deals that took place during the Waxman-Markey and Kerry-Graham-Lieberman days, the maxim that comes to mind is: "No man survives when freedom fails, the best men rot in filthy jails, And those who cry, 'appease, appease' Are hanged by those they tried to please" (Hiram Mann).

I remember so well the Hillary healthcare push of 1993—it was the first attempt to impose socialized medicine. As I was flying back from D.C. through Chicago, I was so pleased we had virtually won the fight and defeated Hillary Care. I was on the plane reading the *Wall Street Journal* where I saw a full page ad by the American Medical Association embracing Hillary Care. Needless to say, I called the AMA from Chicago. They were appeasing the enemy.

And the same thing was happening with many of the energy providers during the cap and trade push.

Democrats were always so pleased with themselves when the corporate heavyweights would come to the table, but I maintained that there was no way they were doing this out of the goodness of their hearts or because they wanted to be good stewards of the environment. They were in this because they knew that they could profit if they played their cards right—and the American consumer would have to foot the bill.

During the Waxman-Markey days, many of the utility companies thought cap and trade was inevitable with Democrats controlling both houses of Congress and the White House. So instead of fighting against the bill because it was bad for jobs and the economy, they started negotiating with the enemy: they were appeasers. But as what always happens to the appeasers, they are, as the saying goes, hanged by those they try to please. And of course, that's exactly what happened in the deals that were made with Waxman-Markey.

One of the reasons they wanted to negotiate was because they felt that regulations under legislation would be at least somewhat better than regulations by the EPA under the Clean Air Act. They also wanted to get a good deal on the allocation of carbon credits, which were limited. But in the end, the electric utilities didn't get what they wanted in the Waxman-Markey bill—it sold the utilities short. And even worse, the Waxman-Markey bill did not fully take away the threat of EPA regulations. It did have language to restrict EPA's ability to set National Ambient Air Quality Standards for greenhouse gases but this restriction was on a limited time frame so it left the door open for both legislation and EPA regulations.

When various companies came to the Senate to shop amendments, I said the last thing I wanted to do was improve the bill. I wanted to kill the bill, so I wasn't much use to them.

Especially during Kerry's last cap and trade push, many of the utilities, which primarily used natural gas and nuclear energy as opposed to coal, saw cap and trade as inevitable. When they realized that coal was getting some extra provisions in the form of a mill tax to support clean coal technologies, they wanted to carve out as many allowances as they could. Now no one is a bigger supporter of natural gas and nuclear energy than I am; I truly believe we need it all, so I told them that coal may be on the chopping block now—as it was the industry that cap and trade proponents' ultimately sought to destroy—but you'll be next.

So when the next Kerry-Graham-Lieberman proposal came along, I asked these appeasers: do you think a cap and trade bill is good for the economy, good for your members, good for workers, good for consumers?

Don't forget what happened with Waxman-Markey: some utilities thought they had a deal, but when the language was actually drafted, the deal made Waxman and Markey happy, but not the utilities.

During the Kerry-Graham-Lieberman debate, one of the biggest appeasers, as many will be surprised to realize was an oil company, BP. The trio had what they thought was a clever plan to get Republicans over to their side: they were going to put in provisions that would encourage offshore drilling for oil and gas, increase nuclear power capacity, and encourage clean coal technologies—and the promise of deals would bring industry to the table. It was a classic divide and conquer strategy—a strategy that was destined for failure because whenever you get votes on one side by using these tactics, you always lose votes on the other side. Besides, even if you were for offshore drilling, you still couldn't get away from the fact that any cap and trade bill would just mean more dependence on foreign oil, more taxes, and fewer jobs. At the same time, President Obama's FY budget planned to impose well over $40 billion in new taxes on the oil and gas industry, which would only discourage production.

The worst part about Kerry-Graham-Lieberman was that on top of the higher electricity prices it would force on consumers, it also contained a gas tax. This gas tax was supported by none other than BP and it was a great deal for them:[321] the American consumer, not the company, would be footing the bill for this greenhouse gas regime—the same consumers were suffering from high unemployment. A new tax was the last thing they needed. When it became clear that the bill contained a gas tax, the Kerry-Graham-Lieberman house of cards came tumbling down. The White House came out opposing it and blamed the idea on Senator Graham.[322] The sponsors and supporters of the bill were left trying to explain how legislation that raises the cost of gasoline for consumers, isn't really a tax.

On April 19, 2010, I spoke with Jeanne Cummings of *Politico* about Kerry, Graham, and Lieberman's latest cap and trade effort.[323] Jeanne specifically focused on the Democrats' aggressive effort to bring the energy development industries to the table and asked if the trio will be successful this time given that the deals they are offering were more generous now than ever. I said that their strategy is flawed. They've tried it before and it didn't work: it's called "divide and conquer" when they go to the various oil, gas, nuclear, and coal industries and carve out special deals. I said that the problem with that is you may be able to get some votes that way but you will lose votes on the other side. At that time, ten Democrats had already signed a letter telling the bill's sponsors that if they add an offshore drilling provision they will no longer support the bill. So with offshore drilling, they might pick up four votes on one side but they lose ten on the other. I said that it would be something that they'll introduce on Earth Day and have a big celebration and then it will fade away quickly.

Only a few days after my interview with Jeanne Cummings, BP, the company that was supposed to help make the bill successful, was the very company responsible for the oil spill that began in April 2010. Not surprisingly, the provisions planned for expanding offshore drilling were immediately abandoned and then an interesting shift in rhetoric took place: whereas initially increasing offshore drilling was going to be a tool to help pass the bill, after the spill, the message was that we had to pass this bill to stop offshore drilling altogether.

In the wake of the spill, President Obama announced that now was the time to put a price on carbon.[324] People had died, people's economic livelihoods were at stake, and the environment was being harmed, yet the President was taking this opportunity to push cap and trade. That's when President Obama began his moratorium on deepwater drilling—something that environmental groups have sought for years. It was an exercise in overreach that would do far more harm than good. The Louisiana Department of Economic Development was estimating that thousands of good paying jobs would be killed in the moratorium's wake. It was a time when we should have been concentrating on putting a cap on the spill, not putting a cap on the economy.

After the drilling provisions were taken out, Senator Graham could no longer support the bill—so their divide and conquer strategy even lost them a former sponsor. The bill that was, from then on, known as Kerry-Lieberman was well on its way to failure.

Whereas Jeanne Cummings of *Politico* wanted to know if cap and trade had a better shot with the drilling provisions, Stuart Varney of Fox News, in an interview in May 2010, wanted to know if the bill had a better shot now that drilling provisions had been taken out. I said no, it is still dead and for the same reasons. Without the drilling provisions, they would just lose those members who supported offshore drilling and they'd be right back where they started.

The reaction from the environmental community to the spill in the Gulf was not unlike their reaction during the time of the *Exxon-Valdez* incident in 1989. At that time, I was serving on two committees in the U.S. House of Representatives that investigated the oil spill. Four days after the tragedy, I spoke to the U.S. Chamber of Commerce in Washington and was taken aback by environmental activists who were celebrating. They were saying that they were glad the accident happened because they could parley this tragedy into stopping development and exploration of the Arctic National Wildlife Refuge. The same thing was happening with the BP oil spill: they were trying to parley this into killing all kinds of offshore drilling. Of course, the *Exxon-Valdez* incident was a transportation accident, and decreasing offshore production would only increase our dependence on foreign oil, which would increase transportation of oil and therefore increase the possibility of another oil spill. But that wasn't the point—they just want to stop drilling altogether.

To all those who still after all this time thought that cap and trade was inevitable, I reminded them that opposition has only grown stronger and more intense. They could have all the backroom secret deals they wanted to try to get to sixty votes, but try as they might, it was never going work. If we look back on the votes in the Senate, it is overwhelmingly clear that support for cap and trade over that time has dropped considerably. In 2003, they got forty-three; in 2005, they got thirty-eight; and in 2008, with Lieberman-Warner, they got forty-eight. But let's not forget that just after the cloture vote on Lieberman-Warner, ten Democrats, nine of whom voted for cloture, very quickly sent a letter stating that they could not vote for Lieberman-Warner "in its current form." So subtract nine, and you get thirty-nine votes. That's a far cry from sixty. There was no way Kerry-Lieberman was going to pass.

CAP AND TRADE IS DEAD BUT THE ENDANGERMENT FINDING IS ALIVE AND WELL

Later in June, even as Kerry and Lieberman were still trying to dig cap and trade out of the grave, the attention began to shift to stopping EPA's greenhouse gas regulations from taking effect.

Here's where the idea of playing "Russian roulette" with regulations, as the *Wall Street Journal* brilliantly pointed out a year earlier, comes in. Democrats' mantra was the EPA regulations would be so much worse and much more expensive than legislation. We had managed to kill cap and trade handily, so now Democrats were going to be responsible for some of the messiest, most expensive and most onerous regulations in American history. With elections coming up, many of them were having second thoughts.

As I mentioned before, one of the reasons EPA regulation of the greenhouse gases would be such a disaster is that the Clean Air Act contains very specific emissions thresholds for regulated pollutants. Under a program to maintain air quality, facilities that emit 250 tons or more per year of a given pollutant must obtain a Prevention of Significant Deterioration, or PSD, permit before they can build or make major modifications to existing facilities. Two hundred and fifty tons is a big number for traditional pollutants such as sulfur dioxide or nitrogen oxide, but not for greenhouse gases. A large commercial building, for example, emits about 100,000 tons of CO_2 a year. We're talking about six million sources potentially subject to EPA regulation. To get around this unmitigated administrative and economic disaster, EPA just decided to ignore the law by instituting the "tailoring rule." That's right: it randomly decreed that regulations would apply only to facilities that emit more than 100,000 tons. That threshold would be tweaked over time and apply to sources at differing stages. But the Clean Air Act is clear. EPA can't just change a law clearly laid out by Congress. Two hundred and fifty tons is two hundred and fifty tons.

What will be the results of this? Imagine heading to church on Sunday to find the doors locked because it couldn't afford to install Best Available Control Technology to reduce its greenhouse gas emissions. Of course, EPA dismisses this and similar examples as nothing more than empty scare tactics. They contend that they have already exempted smaller entities through the tailoring rule so no one has to worry.

Not so fast. EPA's so-called tailoring rule is now being challenged with several lawsuits. It is very likely that the D.C. Circuit will overturn it and force EPA to grapple with the regulatory nightmare of its own creation. If the tailoring rule is thrown out—and almost everything is regulated by the EPA including farms, churches, coffee shops, and restaurants—what will be the economic impacts? According to EPA's own documents, PSD permits cost an average of $125,120 and impose a burden of 866 hours on the applicant. In addition, the nation's largest employers, such as refineries, electric utilities, and industrial manufacturing facilities, will be forced to install (currently undefined) best available control technology (BACT) at their plants to reduce CO_2. EPA has also admitted that if the tailoring rule does not hold up in court, they will have to hire 230,000 new employees and spend an additional $21 billion to implement their greenhouse gas regime.[325]

THE MURKOWSKI RESOLUTION AND ROCKEFELLER COVER VOTES

In June of 2010, Senator Murkowski introduced a resolution that would prevent the EPA from regulating greenhouse gases under the authority of the Clean Air Act. The resolution would allow Congress to overturn regulations from the executive branch by gaining a majority in both the House and the Senate.

At a press conference to discuss the resolution, several Republicans came together in opposition to EPA's greenhouse gas regime, even Senator Graham who still believed in his heart that man-made greenhouse gases were leading to catastrophe. I was on the other end of the spectrum as the one who said that man-made catastrophic global warming is the greatest hoax ever perpetrated on the American people. Everyone else who was standing up there with me and Senator Graham were somewhere in between. But we all agreed on one thing: EPA regulation of greenhouse gases would be a huge disaster. It would hand the agency the greatest regulatory power in history.

When Senator Murkowski first brought up the idea of a resolution, it seemed unlikely that she would reach a majority in a Democrat run Senate. But later it became clear that a lot of members who have elections in 2010 or 2012 wouldn't want to go back to their constituents and say "Look at me. Aren't you proud? I allowed the most massive government take-over of every aspect of our economy to take place"—and this is not to mention all for nothing as it would have no impact on the climate.

Of course, after the oil spill, the Democrat talking point was that the Murkowski resolution was a "Big Oil bailout" that will allow oil companies such as BP to pollute the air. They were exploiting the tragedy in the Gulf to advance a political agenda rooted in the belief that fossil fuels are a destructive nuisance that must be eradicated. And it sees an unrelenting bureaucracy and regulation as the means of realizing a future without them. But that belief, if carried out in the form of EPA's impending greenhouse gas regulatory regime, will mean a radical change to our way of life— a change that will mean fewer jobs, fewer American businesses, higher taxes, energy rationing, and more control of people's lives by a massive, unforgiving bureaucracy in Washington. As with healthcare, the Obama Administration, through the endangerment finding, was and still is on the verge of taking over yet another facet of our economy. Murkowski's resolution presented a fundamental challenge to that view, as it would have prevented EPA from realizing their radical agenda.

Democrats were clearly in a difficult position: if they overturned EPA greenhouse gas regulations, it would be a huge blow to President Obama who would be forced to veto a resolution coming out of a Democrat controlled Senate; but if they allowed them to go unchallenged, many Democrats worried that they would lose their jobs. In an effort to avoid an unpleasant situation for the President, Senator Reid promised moderate Democrats that they would have the opportunity to vote on a bill sponsored by Senator Rockefeller, which would provide a two-year delay of EPA's regulations if they agreed to vote against Murkowski's resolution. On June 9, an article in *The Hill* explained how things were panning out:

> Democratic leaders are scrambling to prevent the Senate from delivering a stinging slap to President Barack Obama on climate change. They have offered a vote on a bill they dislike in the hopes of avoiding a loss on legislation Obama hates. The president is threatening to veto a resolution from Sen. Lisa Murkowski (R-Alaska) that would ban the Environmental Protection Agency (EPA) from regulating carbon emissions. But if the president were forced to use his veto to prevent legislation emerging from a Congress in which his own party enjoys substantial majorities, it would be a humiliation for him and for Democrats on Capitol Hill. So Senate Majority Leader Harry Reid (Nev.) and other Democratic leaders are doing what they can to stop it. They are floating the possibility of voting on an alternative measure from Sen. Jay Rockefeller, a Democrat from the coal state of West Virginia, which they previously refused to grant floor time, Senate sources say.[326]

Seven Democrats took the deal offered to them by Senator Reid and did not vote for the Murkowski resolution. The vote was on June 10, 2010, and it was defeated by 47–53. If these seven senators had voted for the motion to proceed to Senator Murkowski's resolution, the motion to proceed not only would have passed, it would have passed handily. The whole plan was transparent: Democratic leaders, in order to ensure that EPA can micromanage farms and other sources, had to develop a scheme to give cover to Democrat members who opposed the EPA takeover. Those seven members were clearly conflicted. They understood the economic harm that an unfettered EPA bureaucracy could mean for their constituents: fewer jobs, more regulations, higher taxes, and a slower economy. But they were pressured by the president and the base of the Democratic Party—they were warned against defying the president on one of his top initiatives. So they turned to the Rockefeller bill as an alternative. It was the two-year delay of Rockefeller—rather than overturning the endangerment finding—that seemed more politically acceptable.

On the Republican side, every Republican Senator voted for the Murkowski Resolution, which means that the Senate Democrats are solely responsible for whatever regulations the EPA implements. Every lost job, every closed factory, every increase in utility bills or gasoline prices due to the EPA's greenhouse gas regulations are the responsibility of Senate Democrats.

The defeat of the Murkowski resolution wasn't the end of the road. I said that if we ever get a vote on the Rockefeller bill, I trust these seven members—and possibly others who voted "no" on the motion to proceed to Murkowski—will vote with their constituents for Rockefeller and against EPA taking jobs, businesses, and energy out of our struggling economy. Even with the promise a vote on his two-year delay, Rockefeller *voted for* the Murkowski resolution. Interestingly the vote on Rockefeller didn't come up until the next Congress when we forced Reid to face the problem again.

"LAST CALL": 2010 ELECTIONS

With the 2010 election looming, the supporters of cap and trade knew that Republicans were going to make significant gains in the next Congress and time was running out. As Senator Kerry put it, it was the "last call" to pass cap and trade.

The town hall meetings in 2009 and 2010 were not just about the passage of Obamacare; they were also about cap and trade. In fact, cap and trade is one of the main reasons that Democrats had so many losses in the House. The 2010 elections also brought an interesting reversal in campaign strategies for Republicans. While in the 2008 elections, few candidates would dare to question the science behind man-made catastrophic global warming, in 2010, many Republicans not only campaigned against cap and trade but also established themselves strongly as skeptics. Americans voted the global warming advocates out and the skeptics in. And the same thing is happening as the 2012 elections approach.

With the Murkowski Resolution, the excuse of many Senators was that even if they voted to rein in the EPA, the resolution would die in a Democrat-led House. After the 2010 elections, they could no longer use this excuse.

UPTON-INHOFE: THE ENERGY TAX PREVENTION ACT OF 2011

In March of 2011, I introduced S.482, the Energy Tax Prevention Act of 2011, along with Congressman Fred Upton, Chairman of the Energy and Commerce Committee, who has introduced the same bill in the House. Like the Murkowski Resolution, the Energy Tax Prevention Act stops the Obama EPA's backdoor cap and trade regulations from taking effect: it protects jobs in America's manufacturing sector; protects consumers from higher energy costs; puts Congress in charge of the nation's climate change policies; and ensures that the public health provisions of the Clean Air Act are preserved.

On February 9, 2011, I was privileged to testify to the House Subcommittee on Energy and Power on the Energy Tax Prevention Act of 2011. Just as Democrats called the Murkowski resolution the "Big Oil Bailout," at the hearing, Representative Henry Waxman (D-California) said that our bill should be called the "Big Polluter Protection Act."[327] He went on to say that it would "repeal the only authority the administration has to protect our health and the environment without providing any alternative."[328] Here we go again. Contrary to Representative Waxman's claims, our bill leaves all the essential provisions of the Clean Air Act intact. It simply prevents the EPA from regulating greenhouse gases which are not harmful to human health. Imposing energy taxes through EPA's cap and trade regulations and blocking economic development won't make Americans healthier—it will only mean fewer jobs, a higher cost of living, and less growth and innovation.

Amy Harder of the *National Journal* in an article called "Reid Might Be Forced to Deal with His EPA Problem," wrote:

It wasn't supposed to go this way if you're a Democrat.

The 111th Congress was supposed to enact comprehensive climate-change legislation as President Obama prodded action with looming carbon rules.

Instead, efforts to price greenhouse-gas emissions collapsed in the Senate last year and left Obama with his hands tied around his climate-change regulations.

Obama continues to publicly support the regulations and has punted the debate down Pennsylvania Avenue to Congress.[329]

In April 2011, Senate Minority Leader Mitch McConnell offered the Energy Tax Prevention Act as an amendment, which finally forced Reid into action on allowing a vote on the Rockefeller bill—a two year delay—which was also offered as an amendment. Senator Max Baucus also joined in the "cover vote" fray and offered an amendment that stipulates that only the industries would be regulated and not farms and ranches.

The vote on these amendments was a moment of truth for Obama's job-killing greenhouse gas regulations: sixty-four senators voted that day, in various ways, against EPA's cap and trade agenda. Each one chose whether to take the "cover vote" or actually to vote with their constituents for the only real solution to the problem: the Energy Tax Prevention Act.

The other amendments only delayed or put some limitations on how EPA can regulate.

During the debate, most members publicly aligned themselves with concerned constituents, especially manufacturers or farmers, who oppose EPA's greenhouse gas regulations. Democratic Senator Sherrod Brown, for example, wrote a letter to EPA Administrator Lisa Jackson in February, arguing that "any approach to reducing greenhouse gas emissions must recognize the unique situation of energy-intensive manufacturers." Of course, EPA's regulations don't, and can't: "It is disconcerting," the senator wrote, "that, to my knowledge, the EPA has neither a plan in place nor the authority to provide these protections to U.S. manufacturing, a sector of the economy critical to the continued economic recovery of my state and so many others."[330]

Well put. Yet Senator Brown voted against the Energy Tax Prevention Act, the only solution that would fully address the aforementioned concerns. Delays, carve-outs, and exemptions won't solve the underlying problem: now he and all the others who did not vote to stop EPA will have to explain why they stood by and let it happen.

I was very pleased that the House handily passed the Upton-Inhofe bill with overwhelming bipartisan support. In the Senate, we are still ten votes short, but with sixty-four members sending a clear message to the administration that the Obama-EPA needs to be reined in, we will continue fighting. At present, the Senate Democrats have a majority of three, but there are twenty-three Democratic senators up for re-election in 2012. I call eleven of those Senators an "endangered species." Several have announced they will retire rather than run again. I fully expect to be Chairman of the Environment and Public Works Committee again and I hope it may just be a matter of time until we can declare final victory, by stopping EPA from regulating greenhouse gases.

"CAP AND TRADE IS DEAD. LONG LIVE CAP AND TRADE"

In November of 2010, they finally admitted what I had been saying since 2003: Senator Kerry conceded that cap and trade is dead and Senator Lieberman said "whether we like it or not, cap and trade has no chance of passage in the next Congress…And so we've got to find separate ways to go at it."[331]

So, just as Lieberman said, they have since been trying to find separate ways to go at it—that is, to put a price on carbon. One of these ways has been to claim that America has been falling behind other nations in the "clean energy race." That was exactly the message in President Obama's State of the Union address in January 2011. He declared that this is our "Sputnik moment" and that the same progress that we achieved in space technologies can be achieved if we invest in clean energy technologies:

> This is our generation's Sputnik moment. Two years ago, I said that we needed to reach a level of research and development we haven't seen since the height of the Space Race. And in a few weeks, I will be sending a budget to Congress that helps us meet that goal. We'll invest in biomedical research, information technology, and especially clean energy technology—an investment that will strengthen our security, protect our planet, and create countless new jobs for our people.[332]

In this speech, the president set the goal of having 80 percent of America's electricity come from clean energy sources by 2035 and it later became clear that he wanted to do this by establishing a "Clean Energy Standard" (CES). Under a CES, utilities would be required to trade and sell clean energy credits, increasing the mandate required until 80 percent of electricity needs are met by clean energy. As is explained in the Obama Administration *Economic Report*:

> To meet this goal, the Administration is proposing a Clean Energy Standard (CES) that would require electric utilities to obtain an increasing share of delivered electricity from clean sources—starting at the current level of 40 percent and doubling over the next 25 years. Electricity generators would receive credits for each megawatt-hour of clean energy generated; utilities with more credits than needed to meet the standard could sell the credits to other utilities or bank them for future use.[333]

This plan was just cap and trade turned inside out, and it was so short lived that it barely even made the political radar before it was dead—in fact, it was dead on arrival in Congress. Kim Strassel of the *Wall Street Journal* put it well when she wrote just days after the President's speech:

Listen carefully to Mr. Obama's speech and you realize he spent plenty of it on carbon controls. He just used a different vocabulary. If the president can't get carbon restrictions via cap and trade, he'll get them instead with his new proposal for a "clean energy" standard. Clean energy, after all, sounds better to the public ear, and he might just be able to lure, or snooker, some Republicans into going along.[334]

Of course, President Obama had been pushing this "Sputnik moment" since the beginning. In his 2009 State of the Union address, he held up China as an example of how other countries are taking greater strides than the United States on clean energy:

We know the country that harnesses the power of clean, renewable energy will lead the 21st century. And yet, it is China that has launched the largest effort in history to make their economy energy efficient … Well I do not accept a future where the jobs and industries of tomorrow take root beyond our borders—and I know you don't either. It is time for America to lead again.[335]

In December 2010, my Environment and Public Works Committee staff released a report that showed that the so-called "clean energy race" between the United States and China—and the lament that America is losing—is an idea concocted by activists to promote cap and trade, renewable energy mandates, and greater government control of the economy.[336] It is premised on a biased, narrow picture of China's energy development and the demonstrably false notion that economic growth and innovation are best realized through government mandates. If China is embracing anything, it is the reality that fossil fuels, along with nuclear power, are the engines of economic growth and prosperity.

Although the CES was gaining no traction whatsoever, President Obama announced at Penn State that as part of his clean energy push, "So you show us the best ideas to change your game on the ground; we'll show you the money. We will show you the money."[337]

Well the solar company, Solyndra, was one of those companies that was shown the money at the beginning of Obama's presidency, as it was a recipient of the stimulus funds under the American Recovery and Reinvestment Act. At the time, Solyndra was touted as the model to follow by the Administration for its green energy economy, and it was hailed as one of the Recovery Act's biggest success stories. As President Obama said in a speech:

> When it's completed in a few months, Solyndra expects to hire a thousand workers to manufacture solar panels and sell them across America and around the world [...] It's here that companies like Solyndra are leading the way toward a brighter and more prosperous future.[338]

Vice President Joe Biden said that the loan guarantee to Solyndra was an "unprecedented investment this Administration is making in renewable energy and exactly what the Recovery Act is all about" and Energy Secretary Steven Chu said that it was part of "a broad, aggressive effort to spark a new industrial revolution that will put Americans to work."[339]

Now as the country struggles with unemployment and Solyndra has completely collapsed and gone bankrupt, it is clear that the President's policies of "showing the money" has meant throwing the money away. In the end, Solyndra is more than just a bankrupt company: it is a metaphor for the failure of Obama's war on affordable energy and American fossil fuel jobs.

A REAL SOLUTION TO JOBS AND ENERGY SECURITY

In March 2011, in the midst of this "Sputnik moment" clean energy economy push that was clearly failing, the Congressional Research Service report that first surfaced in 2009 was updated and it shows us again that America's combined recoverable natural gas, oil, and coal endowment is the largest on Earth. In fact, our recoverable resources are far larger than those of Saudi Arabia, China, and Canada combined.

While the Obama Administration goes forward with a conscious policy choice to raise energy prices, accomplished in good measure by restricting access to domestic energy supplies, we find out that those supplies are, according to the Congressional Research Service, the largest on Earth.

Here's what CRS says about America's tremendous resource base:

Oil

CRS offers a more accurate reflection of America's substantial oil resources. While America is often depicted as possessing just 2 or 3 percent of the world's oil—a figure which narrowly relies on America's proven reserves of just 28 billion barrels—CRS has compiled U.S. government estimates which show that America, the world's third-largest oil producer, is endowed with 163 billion barrels of recoverable oil. That's enough oil to maintain America's current rates of production and replace imports from the Persian Gulf for more than fifty years.

Natural Gas

Further, CRS notes the 2009 assessment from the Potential Gas Committee, which estimates America's future supply of natural gas is 2,047 trillion cubic feet (TCF)—an increase of more than 25 percent just since the Committee's 2006 estimate. At today's rate of use, this is enough natural gas to meet American demand for ninety years.

Coal

The report also shows that America is number one in coal resources, accounting for more than 28 percent of the world's coal. Russia, China, and India are in a distant second, third, and fifth, respectively. In fact, CRS cites America's recoverable coal reserves to be 262 billion short tons. For perspective, the United States consumes just 1.2 billion short tons of coal per year. And though portions of this resource may not be accessible or economically recoverable today, these estimates could ultimately prove to be conservative. As CRS states: "...U.S. coal resource estimates do not include some potentially massive deposits of coal that exist in northwestern Alaska. These currently inaccessible coal deposits have been estimated to be more than 3,200 billion short tons of coal."

Oil Shale

While several pilot projects are underway to prove oil shale's future commercial viability, the Green River Formation located within Colorado, Wyoming, and Utah contains the equivalent of 6 trillion barrels of oil. The Department of Energy estimates that, of this 6 trillion, approximately 1.38 trillion barrels are potentially recoverable. That's equivalent to more than five times the conventional oil reserves of Saudi Arabia.

America's newly tapped shale deposits, such as the Marcellus in Pennsylvania, the Barnett in Texas, the Haynesville in Louisiana, and the Woodford in Oklahoma have added hundreds of trillions of cubic feet of natural gas to our recoverable resource base. Recent estimates of the Bakken shale suggest that North Dakota could become second only to Texas as the nation's largest producer of crude. And thanks to the Canadian oil sands, the EIA has recently estimated that our neighbors to the north posses a mammoth 178 billion barrels of oil reserves—second only to those of Saudi Arabia.

Methane Hydrates

Although not yet commercially feasible, methane hydrates, according to the Department of Energy, possess energy content that is "immense...possibly exceeding the combined energy content of all other known fossil fuels." While estimates vary significantly, the United States Geological Survey (USGS) recently testified that: "the mean in-place gas hydrate resource for the entire United States is estimated to be 320,000 TCF of gas." For perspective, if just 3 percent of this resource can be commercialized in the years ahead, at current rates of consumption, that level of supply would be enough to provide America's natural gas for more than four hundred years.

Instead of continuing down the failed path of this job-killing global warming green agenda, we could help bring affordable energy to consumers, create new jobs, and grow the economy if the Obama Administration would simply get out of the way so America can realize its true energy potential.[340]

CAP AND TRADE IS DEAD

Even with the final demise of the Kerry-Lieberman bill, it is important to remember that the fight is not over: Moveon.org, the Hollywood elite and the anti-sovereignty internationalists are still out there and working overtime to implement their agenda. But as far as cap and trade as *legislation* is concerned, Stuart Varney put it well in our May 12, 2010, interview when he said, after I told him that Kerry-Lieberman didn't stand a chance, "There you have it: Last word on that: It's dead."

EPILOGUE

GETTING OUR COUNTRY
BACK ON TRACK

O N JULY 1, 2011, one of the headlines in the *Tulsa World* read,
"Inhofe believes swimming in Grand Lake cause of his illness."[341] In fact, I was sure of it: I saw green algae in Grand
Lake earlier that week while I was swimming and a few days later, when I
was back in Washington, I became deathly ill and had to fly back to Tulsa
to recover. I joked about what the headlines would be the next day: "The
environment strikes back" or "Inhofe attacked by the environment at last."

The Sierra Club's reaction was great: they sent me a rose and a get-
well-soon card with the message, "We hope you have a speedy recovery

Reprinted with permission.

and that we can work together to ensure all of our nation's lakes are safe for swimming, drinking and fishing."[342] I really appreciated their humor and even brought the card to show everyone at our next Environment and Public Works Committee hearing.

The jokes continued when I had to miss a big conference on climate change sponsored by the Heartland Institute due to my algae illness. As Stephen Stromberg of *The Washington Post* wrote in a piece called "A Funny Thing Happened at the Climate Denier Conference."

> This is one of the most unintentionally hilarious turns of phrase I've seen in a while. On Thursday, Sen. Jim Inhofe (R-Okla.) was supposed to deliver the opening keynote address at the Heartland Institute's sixth International Conference on Climate Change, a conclave committed to "abandoning the failed hypothesis of man-made climate change." That is, until he sent this statement to the conference's organizers:
>
> "I am sorry that I will not be able to join you today at the Heartland Institute's sixth International Conference on Climate Change. Unfortunately, I am under the weather, but I did want to send a short note to say thank you for all of your hard work and dedication."
>
> That must be some extreme weather.[343]

It was indeed extreme—I don't think I've ever been so sick, but not too sick not to address the conference from afar. I reminded them that the last time this conference was held was just weeks after the House of Representatives passed the Waxman-Markey global warming cap and trade bill. With an overwhelming Democratic majority in the Senate, many predicted that the bill would sail through the Senate and be signed into law by President Obama in time for the UN Climate Conference in Copenhagen.

But, we succeeded in defeating the bill by exposing the huge costs that would be imposed on the American people for no environmental gain. I said that Senate Democrats would not be able to go back to their constituents and say aren't you proud of me? I just voted for the largest tax increase in American history. Further undermining their effort was the latest science which showed that there was no "consensus" on global warming.

I had a similar message for the UN climate conference held in Cancun in December 2010, which I delivered via a YouTube address from Washington. I said, "The fact is, nothing is going to happen in Cancun this year and everyone knows it. I couldn't be happier and poor Al Gore couldn't be more upset: it has been widely reported that he is 'depressed' about Cancun."[344]

CLIMATEGATE 2.0: CRISIS OF CONFIDENCE IN THE IPCC CONTINUES

Of course, nothing happened at the UN global warming conference in December 2011 in Durban, South Africa, either.

But true to form, just in time for the conference, the IPCC was at it again, with the release of the *Summary for Policymakers* for its latest report on extreme weather and global warming. But this time it faced an increasingly skeptical public. In 2007, the IPCC's so-called "smoking gun" report, which claimed that there was an unequivocal link between humans and catastrophic global warming, was all anyone ever talked about. On November 18, 2011, when the discredited IPCC released its *Summary for Policymakers*, nobody except me and Representative Markey even noticed.

If the IPCC wasn't discredited enough, another batch of Climategate emails, now known as Climategate 2.0, also surfaced on November 22, 2011, just weeks before the UN climate conference in December. These emails show more of the same manipulation of data and politicization of the science by scientists contributing to the IPCC that was revealed in the original Climategate scandal. One particular email with the title "inhofe & mann & me" from journalist Andrew Revkin to Stefan Rahmstorf, a lead author of the IPCC Fourth Assessment Report, shows an interesting behind-the-scenes take on how they felt about my message. As Rahmstorf writes, "Hi Andy, from over here, it is hard to see this kind of Inhofe speach [sic] as anything else than an irrelevant piece of absurd theatre. It doesn't even bother me any more—he's simply lost it. Cheers, Stefan"

This email was written on September 26, 2006, and the speech to which Rahmstorf is referring is my speech on the "Hot and Cold Media Spin Cycle" which called out the media for using global warming hysteria to sell news, and challenged them instead to take an objective approach to the science.

Revkin replies, "I know, but he still speaks to and for a big chunk of America—people whose understanding of science and engagement with such issues is so slight that they happily sit in pre-conceived positions."

If Revkin is right about one thing in this email, it is that I speak to and for a big chunk of America. Remember after I delivered that speech in 2006, my Committee office was inundated with calls from Americans thanking me for having the voice of reason amid all the hysteria. But my ability to reach a wide audience is not due to Americans having only a "slight" understanding of the science as Revkin alleges. It is due to the fact that American people understand all too well the politicization of the science, the manipulation of the data, and the "tricks" implemented by many IPCC scientists to come to preconceived solutions.

The IPCC can only blame itself for its irrelevance today and everyone knows it, including many in the liberal media. As Joe Romm of Climate Progress said about these emails being released just before the UN climate conference in Durban, "It's so refreshing that anybody thinks those climate talks actually matter."[345]

WHAT WILL THEY TRY NEXT?

Of course, Al Gore and Big Green may be down at the moment but they are not out. These global warming alarmists have not given up their efforts to continue to push their agenda. At the climate conferences in Cancun and Durban, some leaders were stepping up their attacks on capitalism and United Nations officials were saying they need to do more to "spread the wealth around."

It was just more of the same, only now they no longer have global warming hysteria to back it up. So what will they do next to try to implement their green regime?

Looking back, it is crystal clear that this debate was never about saving the world from man-made global warming; it was always about how we live our lives. It was about whether we wanted the United Nations to "level the playing field worldwide" and "redistribute the wealth." It was about government deciding what forms of energy we could use.

From the alarmism to their so-called solutions, the issue has always pitted big government supporters against strong individualism. That mindset has always led me to believe that the liberal agenda pushing the global warming movement will never win. But that doesn't mean they won't keep on trying.

The first big question they will be forced to answer: will they keep Al Gore, Rajendra Pachauri, and James Hansen as the three faces of the movement? While the trio did wonders in catching the media's attention during the glory years of alarmism, each has faced their own troubles ever since. Al Gore's problems are obvious, but there have also been calls for Pachauri to resign as head of the IPCC. Hansen also faces questions from his own side because of his environmental activism and his inconvenient campaign against cap and trade.[346]

The other big question they will be forced to answer is: will they stay on track with their green energy future talking points or go back to try and scare the American people through alarmism again?

They tried alarmism in October 2010, and it failed miserably. A group called 10:10 released a film, *No Pressure,* that featured children who don't believe in global warming being blown up by their teacher. The group said that it was meant to be funny, but it created so much outrage with the public that the video had to be immediately pulled down from the organization's Web site. It was the most outrageous, last-ditch effort to scare little kids into thinking that they could be killed if they don't believe what they're told to believe.

And it was no amateur production team. As the *UK Guardian* reported, the film featured "film star Gillian Anderson and England footballer Peter Crouch, with music donated by Radiohead and shot by a forty-strong professional film crew led by director Dougal Wilson, it was intended to galvanise viewers into taking personal action to reduce their own carbon footprint."[347]

The *Guardian* also writes this line from one of the child actors: "Jamie Glover, the child-actor who plays the part of Philip and gets blown up, has similarly few qualms: 'I was very happy to get blown up to save the world.'"[348]

Then in October 2011, the media had the chance they had been waiting for to promote alarmism again. Richard Muller, a professor at the University of California at Berkeley, released a report from the Berkeley Earth Surface Temperatures project team (BEST) that claimed that the world was warming at an alarming rate—conveniently just in time for the 2011 UN climate change conference in Durban, South Africa.[349] In an op-ed the *Wall Street Journal* Muller boldly declared that the age of skepticism was over and that "you should not be a skeptic, at least not any longer."[350]

The media, which had gone through a prolonged cooling spell on global warming, jumped on the story immediately. AP reporter Seth Borenstein, the same reporter who wrote the glowing article, "Scientists give two thumbs up to Gore's movie" in 2006, wrote an article on Muller's report called "Skeptic finds he now agrees global warming is real."[351] Of course, in typical fashion, a fawning media storm ensued. David Rose wrote in an October 30, 2011, Mail Online article called, "Scientist who said climate change sceptics had been proved wrong accused of hiding truth by colleague" about the media mania that was drummed up, explained that "It was cited uncritically by, among others, reporters and commentators from the BBC, the *Independent*, the *Guardian*, the *Economist,* and numerous media outlets in America. *The Washington Post* said the BEST study had 'settled the climate change debate' and showed *that anyone who remained a skeptic was committing a 'cynical fraud.'"[352]

This indulgence in alarmism, however, was so short lived it was over before many even realized it started. Upon further review, it turned out that Muller's study did not evaluate if the warming was man-made and it also was not properly peer reviewed. In fact, Muller released his findings without even informing his colleagues in the study, Professor Judith Curry and Anthony Watts. As Professor Curry said, "Of course this isn't the end of skepticism. To say that is the biggest mistake he [Prof Muller] has made."[353]

By the way, Muller is far from being a former skeptic. In fact, we specifically excluded him from our Senate report on skeptic scientists because he was clearly on the other side of the debate. We haven't heard about Richard Muller since.

We also haven't heard much about green jobs lately.

With the total collapse of Solyndra and dismal unemployment numbers, their "clean energy future" and "clean energy jobs" talking points have clearly failed as well so it will be interesting to see what they try next.

2012 ELECTIONS: STOPPING OBAMA'S WAR ON AFFORDABLE ENERGY

Today the mood in Washington is significantly different. Everyone readily admits that cap and trade legislation is dead on Capitol Hill—even our good friend, Senator Boxer.[354]

Now as the 2012 elections approach Presidential candidates are dropping global warming like a hot potato. Presidential candidate and former Speaker of the House Newt Gingrich told Fox News on November 8, 2011, that holding hands with Nancy Pelosi on the sofa saying we must do something about global warming in an ad sponsored by Al Gore's climate alliance is "probably the dumbest single thing I've done in recent years."[355]

I couldn't agree more, and I appreciate his honesty.

With the hysteria about catastrophic man-made global warming behind us, our fight against the hoax has shifted to stopping President Obama from imposing through regulation what he was unable to achieve through legislation.

With nineteen House Democrats supporting the Upton-Inhofe Energy Tax Prevention Act, and sixty-four senators on the record in some way against EPA, all eyes are on EPA and the White House. Will EPA change course? Will President Obama accept that his cap and trade agenda is wildly unpopular, and agree to drop it? Don't hold your breath.

Now in the wake of the EPA Inspector General report which reveals that EPA did not engage in the proper record-keeping process or follow the required peer review procedures leading up to the endangerment finding—on top of the fact that the science comes from the IPCC, whose credibility is seriously called into question—the very foundation of the endangerment finding is quickly crumbling, and I think it will only be a matter of time before it collapses.[356]

In the meantime, the debate continues, and the battle over the Energy Tax Prevention Act and Obama's war on affordable energy carries on. The bill will come to the floor again, and soon, so members will once again have to decide whether they stand with consumers, manufacturers, farmers, and small businesses, or with EPA's barrage of greenhouse gas regulations that will harm all of them.

My hope is that if the endangerment finding does not collapse under the weight of its own flaws, which I believe it will, the Energy Tax Prevention Act will have a clear path towards victory after the 2012 elections.

MOVING ON FROM THE HOAX

What I said on the Senate floor in July 2003 is exactly the same message I have had ever since. The science behind man-made catastrophic global warming simply isn't there and the United States Senate would never ratify Kyoto or pass cap and trade. I am vindicated on these points today and now that the global warming hysteria is behind us, its time to get back to powering this amazing machine called America, which is the surest way to revive our ailing economy. The United States has a clear choice: we can continue implementing Al Gore and President Obama's global warming agenda, which is destroying jobs and doing great harm to our economy, or we can elect officials who support developing our nation's vast natural resources, which are the key to our nation's recovery. With the November 2012 elections on the horizon, we finally have the chance to stop Obama's cap and trade agenda and its mindless restrictions on our ability to develop and produce our own resources.

But the demise of the Obama Administration does not mean that they won't keep trying. They'll be back: MoveOn.org, the Hollywood elite, the anti-sovereignty internationalists are still out there and they still have the resources to be dangerous.

However, I firmly believe that when the history of our era is written, future generations will look back and wonder why we spent so much time and effort on global warming and pointless "solutions" like Kyoto and cap and trade. In the end, through all the hysteria, all the fear, and all the phony science, what global warming alarmists have often forgotten is that God is still up there, and as Genesis 8:22 reminds us:

"As long as the earth remains,

there will be springtime and harvest,

cold and heat, winter and summer,

day and night."[357]

AFTERWORD

WHAT GLOBAL WARMING AND EARMARKS HAVE IN COMMON

O R THIS COULD BE ENTITLED, "President Obama's greatest victory...a gift from Republicans."

This will be the most difficult concept for conservatives to comprehend, let alone embrace. As I watched this "gift" unfold, this ceding of our Constitutional authority to President Obama, I felt perhaps the greatest frustration of all in my over ten-year battle against the hoax. How would the Obama bureaucracy be emboldened with their enhanced authority? How could they use it to advance their effort to pass the greatest tax increase in history, cap and trade? As difficult as it is, I will ask you to read on. Difficult? Yes, because it contradicts a basic tenant that most conservatives have been led to believe. However, there is a happy ending.

THE LONE VOICE IN THE WILDERNESS

On November 22, 2010, *Roll Call* published an article entitled, "Inhofe Happy to Stand Apart" which laid out the reasons why I am often alone opposing some key "politically correct" issues—most recently the moratorium on earmarks:

> In an interview last week, [Inhofe] recalled a time when one of his grandchildren "came up to me and said, 'Pop-I, Why do you always do things that nobody else does?' ... and I said, 'because nobody else does.'"
>
> Case in point: As many of his GOP colleagues reversed long-held positions on an earmark ban last week, Inhofe proudly defended his support of the practice.
>
> Republicans adopted an internal rules change to create a voluntary ban on requesting any kind of earmark—including transportation projects, tax cuts for particular industry sectors or other Congressionally directed funding.[358]

Of course, I would so much rather join the crowd and hold hands and have a ban on earmarks and everybody's happy. But I do what nobody else does because I've got twenty kids and grandkids. The things we're doing today are not for me, they're for the future. In the 1970s, I was the "lone voice in the wilderness" in Oklahoma pushing for the balanced budget amendment; I risked expulsion in the House of Representatives to overturn the discharge petition that was allowing members to hide their votes from the public; for many years, I was the only one willing to take the most politically incorrect position at the time and challenge the science behind the global warming hoax; and now I am one of the few senators standing up against the earmark moratorium in Congress. But, as before, I will eventually be proven right.

Before I go any further, let me redeem myself with my conservative readers. The American Conservative Union has ranked me the number one most conservative member in the United States Senate. *Human Events,* in editorializing on the "Top 10 Most Outstanding Conservative Senators," ranked me number one, calling me an "unabashed conservative; he's unafraid to speak his mind." I was also recognized by *The National Journal* as the number one conservative in the United States Senate for 2009.

While I am often the lone voice in the wilderness, I am glad that my good friend Paul Weyrich, co-founder of the Heritage Foundation and a leading conservative, agreed with me in a column for Renew America called "Senator Inhofe: Transportation, Work, and Achievement," which was published September 12, 2006. This piece is very important to me because it was one of the last things Weyrich wrote before his death. As Weyrich rightly said, "[Inhofe] and I always tell our fellow conservatives that the two matters as to which the Federal Government is authorized to spend money are defense and infrastructure." He called me a work horse not a show horse because of my efforts on bills that receive less attention, such as the Water Resources Development Act (WRDA)—an important bill that funds water infrastructure projects. Of course, Weyrich put it best so I'll let him speak to that:

∙∙

Senator Inhofe: Transportation, Work, and Achievement
By Paul Weyrich
September 12, 2006

When I came to work in the United States Senate, 40 years ago this January, I quickly learned that there are two kinds of Senators—work-horses and show horses. I dare say few, if any, high school students could name all 100. Indeed most teachers would be impressed if their high-schoolers could name the two Senators from their own State.

I have watched over the years the Senators who never met a microphone they didn't try to get in front of. Then I have watched the Senators who work quietly on matters vital to the nation but who get very little coverage for doing so.

One of the workhorse Senators is James M. Inhofe (R-OK). His is hardly a household name outside his own state, where he wins by landslide margins. In the Senate he doggedly works on various pieces of non-sexy legislation. Often his work pertains to national defense. I have seen him go toe to toe with both the Clinton and Bush Administrations. And he won. I have seen him clash with the Congressional leadership of his own party. For example, he got the rules changed so that Congressmen who sign a discharge petition (to force a bill to the floor against the wishes of the leadership) must do so in broad daylight. The rules previously permitted them to hide behind procedure.

Having been trained by two workhorse Senators I appreciate them a lot more than those who will say anything to get on television. The reason I mention Jim Inhofe is because of the 100 Senators I would put him as the top workhorse Senator. He works on many projects at once. He pursues them until they are complete. Do not get me wrong, he is good on television. Since the advent of the Fox News Channel, he now has begun to get some exposure and he does well. Primarily, however, he does what he is now doing—that is, working on an infrastructure bill that has almost no national following. He is shepherding something called the Water Resources Development Act (WRDA). He and I always tell our fellow conservatives that the two matters as to which the Federal Government is authorized to spend money are defense and infrastructure. Two summers ago Inhofe secured passage of the Transportation Bill, which took incredible skill on his part. Yes, it has a few questionable items but by and large that bill was an extraordinary piece of work. I praised him for it at the time and I do so again today, despite all the criticism. We both believe that spending outside of defense and infrastructure is stretching the Constitution to a point beyond recognition.

Anyway, back to this legislation, the bill Inhofe is now working on authorizes the Corp of Engineers to do flood control, navigation and environmental restoration projects. For example, the average transportation cost savings of users of the inland waterway system is $10.76 per ton hauled or $7 billion annually over rail, highways and air transportation.

Flood control, as demonstrated during Katrina and Rita, is a critical service provided by the Army Corps of Engineers. Money was appropriated to fix those infamous levies in New Orleans but local politicians always diverted the money to their own projects and now we are all paying the price. Nevertheless, according to the American Society of Civil Engineers, flood control structures on average prevent $22 billion in flood damage per year. That is a saving of $6 for every $1 spent.

Clearly, projects that promote economic growth through good movements or prevent damage due to flooding are not pork. Yet many in the media, who never understand the big picture in this country, pick on some project in a Congressman's district and charge him with bringing home the pork. Not always so. Recognizing that not all proposed flood control or navigation projects are necessary, the Senate has established firm criteria for evaluating project requests.

First, projects have to have a chief report, which means that the Corps of Engineers has determined that the project is technically feasible, environmentally sound and economically justified. Second, Inhofe and his committee attempt to oppose any environmental infrastructure project which is outside the scope of the main mission. You can imagine that there are Senators on Inhofe's committee who do bring pork to the table. Inhofe won't budge on that point. Finally, Inhofe's Environment and Public Works Committee opposes cost waivers, thus following the policy established in the WRDA Bill of 1986 which established cost-sharing requirements. In order for a project to be built local communities must be willing to pay some cost of the project. The same is true in the Transportation Bill only in that measure there is a huge disparity between highways and transit. With highways the Federal Government pays 80 to 90% of the project. With transit, say a light rail line in Denver, the Federal Government will only pay on average around 50%.

Just as in the Transportation Bill (known around here as SAFETEA-LU), in which the Senator got his Committee to agree that projects eligible for Highway Trust Fund dollars be on the State's transportation, the Senate WRDA Bill established and stayed with strict criteria for WRDA projects in an attempt to avoid funding any project which is not justified.

Work on this measure has been long and hard. Inhofe wants to get the final bill passed in these waning days of the 109th Congress. But for Senators like Inhofe (and there are not many—eight or nine at best) who are willing to do the non-exciting, non-sexy work, the real business of the Senate would not go on. The WRDA Bill is important and we can be thankful that Inhofe is behind it, inching it along to enactment.[359]

I am not including the Weyrich op-ed a as self ingratiating gesture but to set up the concept that bureaucratic earmarks, as opposed to congressional earmarks, played a significant role in promoting the whole hoax, using the bureaucracy as a brainwashing mechanism. Read on, you'll see.

LEGISLATORS VS. BUREAUCRATS

There is a very important difference between legislators and bureaucrats.

Legislators are *elected* officials—at the federal level, Congresspersons serving in the United States House of Representatives, and Senators serving in the United States Senate. Legislators pass laws that are signed into effect by the president as the chief administrator of the government. Legislators are periodically re-elected, or un-elected, by their constituencies. They are accountable to the people.

In very sharp contrast, agency bureaucrats are *appointed* officials—they are part of the executive branch of government, ultimately approved by the President. They are not elected. They are not accountable to the people. They serve and are accountable to the president.

The key question for every citizen is: Who do you want making decisions about your life? Someone who is elected that you can vote out of office, or an unelected, unaccountable bureaucrat?

A POWER GRAB BY THE EXECUTIVE BRANCH

Through an overwhelming number of regulations, we are experiencing a masterful power grab on the part of the Executive Branch—in other words, the president.

This is especially obvious in the efforts by the Obama administration to take over regulation of a number of sectors of our society such as healthcare and banking. More recently, the power grab has extended to control of energy, power, global warming, and climate change. Obama's power-grab through the EPA is as massive as that of the Obama healthcare bill.

When I was privileged to testify before a House of Representatives Subcommittee on Energy and Power in February 9, 2011, concerning the "Energy Tax Prevention Act of 2011." I explained the goal of the bill is to keep the EPA from imposing climate-change regulations.

The legislation is designed to reassert the authority of Congress rather than to allow unelected bureaucrats to decide regulations on all forms of energy and to specifically hold the EPA accountable.

Of course, Representative Henry Waxman (D-CA) said that our bill should be called the "Big Polluter Protection Act."[360] He went on to say that our bill would "repeal the only authority the administration has to protect our health and the environment without providing any alternative."[361] However, Congressman Waxman should know there is no "alternative" even necessary because there is no need for the administration to take such economically severe measures to protect our health and environment. There is no *threat* to our health and environment.

What happened to push Congressman Upton and me to action? Since the normal legislative process has failed the Obama Administration due to lack of support from the Congress and the American people, they are pursuing regulations to enact what would amount to a cap-and-tax law. The good news at this point is that the House of Representatives is eager to pass such a change in EPA powers. The Senate isn't quite there yet. I'm eager for that day to come.

THE EPA CLAIMS AND MY REBUTTAL

The EPA claims that it has both justification and Supreme Court authority to regulate CO_2 as a pollutant and to impose regulations that will end up seriously impacting virtually every aspect of American business and daily life.

My leading contention before the subcommittee was that the Clean Air Act—a bill designed to deal with air quality—was passed by Congress with a deliberate intent *not* to regulate so-called greenhouse gases. The Clean Air Act had nothing to do with climate change. Its primary directives were related initially to the regulation of what were called "criteria pollutants" such as ozone, particulate matter, and lead.

Furthermore, the Waxman-Markey bill that the House passed *did* have to do with greenhouse gases. That bill died with Senate inaction and the election of a new Congress. Greenhouse gasses were an issue in the 2010 election that saw control of the House move to the Republicans. I have no doubt that a bill such as Waxman-Markey would not pass the House today.

In her testimony before the subcommittee, EPA Administrator Lisa Jackson referenced the 5–4 decision by the Supreme Court, *Massachusetts v. EPA*,[362] testifying that the Supreme Court concluded that the Clean Air Act's definition of air pollutants includes greenhouse gases.[363]

What the Supreme Court said was this: the EPA has the discretion to decide whether greenhouse gases endanger public health and welfare. However, Administrator Jackson proceeded in making the endangerment finding that greenhouse gases are a threat to human health, and welfare.

THE GREATER TREND TOWARD PRESIDENTIAL RULE

The issues involving the EPA are symptomatic of a much broader escalating trend toward greater and greater Presidential authority through bureaucratic agencies. This trend continues to a great extent, I believe, because people use terminology that hides the process.

In a word...earmarks. I consider it one of the most misunderstood words in our American language today.

WHAT REALLY IS AN EARMARK?

If you ask people today what's wrong with the federal government, they may say runaway deficits or wasteful spending, and I agree with that. However, they may also say earmarks, and I believe they do that because very few people really understand just what an earmark is. In order to understand earmarks, you must know the difference between Congressional authorizations and Congressional appropriations, and you must further understand the difference between Congressional earmarks and Obama Administration earmarks. Yes, there are two types of earmarks.

Congress passes laws that authorize specific programs and either authorizes the overall dollar amount for the program or simply states "such sums as may be necessary." An example of an authorization act is the Energy Independence and Security Act in 2007.[364]

After a law has been passed, Congress then appropriates funds to carry out programs contained in the law, across all of the federal departments and agencies. During the appropriations process, Congress may designate, through an earmark, that a certain amount of the funds available must be spent on a particular program or issue. For example, they may specify that under the Highway program, $100,000 must be used to help a small community deal with a deteriorating highway. This $100,000 is taken out of the larger pool of money that Congress authorizes for the program.

These earmarks or appropriations are not additional funds added to the federal budget.

Here's the crux of the issue: The vast majority of funds are not earmarked by Congress. Executive-branch officials decide how to spend the funds in the vast majority of all programs. The Energy Independence and Security Act of 2007 provides for greenhouse gas and climate research in a variety of sections of the law. The money is spent by either political appointees in the Executive Branch, or in most cases, by career bureaucrats in each of the agencies and departments. But in the final analysis, it's Obama.

One starts to see the connection between global warming and climate research and earmarks. During the decade-long battle against cap and trade and other climate legislation, I was not only fighting the global-warming enthusiasts but also the process that was allowing billions of dollars that were given to *unelected* bureaucrats to dish out as they desired. Keep the concept very clear: unelected bureaucrats are given supervisory control over bureaucratic-agency earmarks.

You might ask why Executive branch or bureaucratic earmarks are any worse than Congressional earmarks. The answer is that in bureaucratic earmarks, most of the decisions are made by the career bureaucrats who are unaccountable to the people.

The people elect their 535 Congressional representatives, the House and the Senate, who have a much better idea of how the people want their tax dollars spent. On the other hand, they elect one president, who is responsible for the entire Executive branch. The Congressional Research Service reports that the president, in turn, appoints approximately one thousand cabinet and subcabinet officials who are confirmed by the Senate,[365] with additional hundreds of political appointees. In all, these bureaucrats run government agencies—fielding requests for grants and other types of funding, submitting those requests to the president for his budget and then, when Congress fails to do its part, receiving jurisdiction over that money to pass it on to those of their choosing.

While having a thousand Senate confirmed "managers" and a few thousand related staffers might seem like enough, you must also keep in mind that these employees of the federal government oversee hundreds of thousands of federal employees who make decisions every day from Washington, D.C., about how your federal tax dollars are spent.

Are you aware that most federal employees serve regardless who is elected president? Just under the political appointees are a group of managers called the Senior Executive Service. These Senior Executives, in turn, oversee at least thousands of additional managers, all of whom have authority to spend federal dollars. Our federal workforce is growing at an alarming rate; the Obama Administration has reportedly added more than 200,000 new positions.[366] The exact number may be disputable. What is not disputable is that these unelected bureaucrats, who do not answer to the public or the voters, make decisions every day on how to spend your tax dollars.

Even if every member of Congress and all one thousand of the president's political appointees try to guard against wasteful spending, they are grossly outnumbered by the career bureaucratic managers and staff.

How does this impact climate change? Quite simply, the career bureaucrats at the EPA, the National Oceanic and Atmospheric Administration, the U.S. Fish and Wildlife Service, and numerous other departments and agencies have been earmarking climate science money to their pet researchers for decades; regardless of the will of Congress or who the president might be. It may shock you to know which presidential administration spent more on climate research. It wasn't Bill Clinton. The Administration of George W. Bush spent more than three times more money on climate research than the Clinton Administration. Who decided how to spend the money? It wasn't President Bush or Vice President Cheney.

Career bureaucrats across the federal government spent hundreds of millions of dollars during the Bush Administration funding the work of Dr. Mann and his cohorts, and it wasn't through Congressional earmarks. A Congressional earmark to fund Dr. Mann would have never passed Congress, not as long as I was in the U.S. Senate.

WHERE DOES THE MONEY GO?

In March 2010, talk show host and political pundit Sean Hannity had a series about what he considered to be the 102 worst "earmarks" related to government spending of tax dollars. He listed these in reverse order from 102 to 1. Read them all. It will make you appreciate the surprise ending at the conclusion of the list.

102. Protecting a Michigan insect collection from other insects ($187,632)

101. Highway beautified by fish art in Washington ($10,000)

100. University studying hookup behavior of female college coeds in New York ($219,000)

99. Police department getting 92 Black Berries for supervisors in Rhode Island ($95,000)

98. Upgrades to seldom-used river cruise boat in Oklahoma ($1.8 million)

97. Precast concrete toilet buildings for Mark Twain National Forest in Montana ($462,000)

96. University studying whether mice become disoriented when they consume alcohol in Florida ($8,408)

95. Foreign bus wheel polishers for California ($259,000)

94. Recovering crab pots lost at sea in Oregon ($700,000)

93. Developing a program to develop "machine-generated humor" in Illinois ($712,883)

92. Colorado museum where stimulus was signed (and already has $90 million in the bank) gets geothermal stimulus grant ($2.6 million)

91. Grant to the Maine Indian Basketmakers Alliance to support the traditional arts apprenticeship program, gathering and festival ($30,000)

90. Studying methamphetamines and the female rat sex drive in Maryland ($30,000)

89. Studying mating decisions of cactus bugs in Florida ($325,394)

88. Studying why deleting a gene can create sex reversal in people, but not in mice in Minnesota ($190,000)

87. College hires director for project on genetic control of sensory hair cell membrane channels in zebra fish in California ($327,337)

86. New jumbo recycling bins with microchips embedded inside to track participation in Ohio ($500,000)

85. Oregon Federal Building's "green" renovation at nearly the price of a brand new building ($133 million)

84. Massachusetts middle school getting money to build a solar array on its roof ($150,000)

83. Road widening that could have been millions of dollars cheaper if Louisiana hadn't opted to replace a bridge that may not have needed replacing ($60 million)

82. Cleanup effort of a Washington nuclear waste site that already got $12 billion from the Department of Energy ($1.9 billion)

81. Six woodlands water taxis getting a new home in Texas ($750,000)

80. Maryland group gets money to develop "real life" stories that underscore job and infrastructure-related research findings ($363,760)

79. Studying social networks, such as Facebook, in North Carolina ($498,000)

78. Eighteen (18) North Carolina teacher coaches to heighten math and reading performance ($4.4 million)

77. Retrofitting light switches with motion sensors for one company in Arizona ($800,000)

76. Removing graffiti along 100 miles of flood-control ditches in California ($837,000)

75. Bicycle lanes, shared lane signs, and bike racks in Pennsylvania ($105,000)

74. Privately-owned steakhouse rehabilitating its restaurant space in Missouri ($75,000)

73. National dinner cruise boat company in Illinois outfitting vessels with surveillance systems to protect against terrorists ($1 million)

72. Producing and transporting peanuts and peanut butter in North Carolina ($900,000)

71. Refurnishing and delivering picnic tables in Iowa ($30,000)

70. Digital television converter box coupon program in D.C. ($650,000)

69. Elevating and relocating 3,000 feet of track for the Napa Valley Wine Train in California ($54 million)

68. Hosting events for Earth Day, the summer solstice, in Minnesota ($50,000)

67. Expanding ocean aquaculture in Hawaii ($99,960)

66. Raising railroad tracks 18 inches in Oregon because the residents of one small town were tired of taking a detour around them ($4.2 million)

65. Professors and employees of Iowa state universities voluntarily taking retirement ($43 million)

64. Minnesota theatre named after Che Guevara putting on "socially conscious" puppet shows ($25,000)

63. Replacing a basketball court lighting system with a more energy efficient one in Arizona ($20,000)

62. Repainting and adding a security camera to one bridge in Oregon ($3.5 million)

61. Missouri bridge project that already was fully funded with state money ($8 million)

60. New hospital parking garage in New York that will employ less people ($19.5 million)

59. University in North Carolina studying why adults with ADHD smoke more ($400,000)

58. Low-income housing residents in one Minnesota city receiving free laptops, WiFi, and iPod Touches to "educate" them in technology ($5 million)

57. University in California sending students to Africa to study why Africans vote the way they do in their elections ($200,000)

56. Researching the impact of air pollution combined with a high-fat diet on obesity development in Ohio ($225,000)

55. Studying how male and female birds care for their offspring and how it compares to how humans care for their children in Oklahoma ($90,000)

54. University in Pennsylvania researching fossils in Argentina (over $1 million)

53. University in Tennessee studying how black holes form (over $1 million)

52. University in Oklahoma sending 3 researchers to Alaska to study grandparents and how they pass on knowledge to younger generations ($1.5 million)

51. Grant application from a Pennsylvania university for a researcher named in the Climategate scandal ($500,000)

Don't give up. There's a reason for this. You're half-way there.

50. Studying the impact of global warming on wild flowers in a Colorado ghost town ($500,000)

49. Bridge built over railroad crossing so 168 Nebraska town residents don't have to wait for the trains to pass ($7 million)

48. Renovating an old hotel into a visitors center in Kentucky ($300,000)

47. Removing overgrown weeds in a Rhode Island park ($250,000)

46. Renovating 5 seldom-used ports of entry on the U.S.-Canada border in Montana ($77 million)

45. Testing how to control private home appliances in Martha's Vineyard, Massachusetts, from an off-site computer ($800,000)

44. Repainting a rarely-used bridge in North Carolina ($3.1 million)

43. Renovating a desolate Wisconsin bridge that averages 10 cars a day ($426,000)

42. Four new buses for New Hampshire ($2 million)

41. Repaving a 1-mile stretch of Atlanta road that had parts of it already repaved in 2007 ($490,000)

40. Florida beauty school tuition ($2.3 million)

39. Extending a bike path to the Minnesota Twins stadium ($500,000)

38. Beautification of Los Angeles' Sunset Boulevard ($1.1 million)

37. Colorado Dragon Boat Festival ($10,000)

36. Developing the next generation of supersonic corporate jets in Maryland that could cost $80 million each ($4.7 million)

35. New spring training facilities for the Arizona Diamondbacks and Colorado Rockies ($30 million)

34. Demolishing 35 old laboratories in New Mexico ($212 million)

33. Putting free WiFi, Internet kiosks, and interactive history lessons in 2 Texas rest stops ($13.8 million)

32. Replacing a single boat motor in a government boat in D.C. ($10,500)

31. Developing the next generation of football gloves in Pennsylvania ($150,000)

30. Pedestrian bridge to nowhere in West Virginia ($80,000)

29. Replacing all signage on 5 miles of road in Rhode Island ($4,403,205)

28. Installing a geothermal energy system to heat the "incredible shrinking mall" in Tennessee ($5 million)

27. University in Minnesota studying how to get the homeless to stop smoking ($230,000)

26. Large woody habitat rehabilitation project in Wisconsin ($16,800)

25. Replacing escalators in the parking garage of one D.C. Metro station ($4.3 million)

24. Building an airstrip in a community most Alaskans have never even heard of ($14,707,949)

23. Bike and pedestrian paths connecting Camden, N.J., to Philadelphia, Pennsylvania, when there's already a bridge that connects them ($23 million)

22. Sending 10 university undergrads each year from North Carolina to Costa Rica to study rain forests ($564,000)

21. Road signs touting stimulus funds at work in Ohio ($1 million)

20. Researching how paying attention improves performance of difficult tasks in Connecticut ($850,000)

19. Kentucky Transportation Department awarding contracts to companies associated with a road contractor accused of bribing the previous state transportation secretary ($24 million)

18. Amtrak losing $32 per passenger nationally, but rewarded with windfall ($1.3 billion)

17. Widening an Arizona interstate even though the company that won the contract has a history of tax fraud and pollution ($21.8 million)

16. Replace existing dumbwaiters in New York ($351,807)

15. Deer underpass in Wyoming ($1,239,693)

14. Arizona universities examining the division of labor in ant colonies (combined $950,000)

13. Fire station without firefighters in Nevada ($2 million)

12. "Clown" theatrical production in Pennsylvania ($25,000)

11. Maryland town gets money but doesn't know what to do with it ($25,000)

10. Investing in nation-wide wind power (but majority of money has gone to foreign companies) ($2 billion)

9. Resurfacing a tennis court in Montana ($50,000)

8. University in Indiana studying why young men do not like to wear condoms ($221,355)

7. Funds for Massachusetts roadway construction to companies that have defrauded taxpayers, polluted the environment, and have paid tens of thousands of dollars in fines for violating workplace safety laws (millions)

6. Sending 11 students and 4 teachers from an Arkansas university to the United Nations climate change convention in Copenhagen, using almost 54,000 pounds of carbon dioxide from air travel alone ($50,000)

5. Storytelling festival in Utah ($15,000)

4. Door mats to the Department of the Army in Texas ($14,675)

3. University of New York researching young adults who drink malt liquor and smoke pot ($389,357)

2. Solar panels for climbing gym in Colorado ($157,800)

1. Grant for one Massachusetts university for "robobees" (miniature flying robot bees) ($2 million)

Grand Total: $4,891,645,229[367]

After Hannity's program aired, I repeated his list in a speech I gave on the Senate Floor. I asked, "What do all of these 102 'earmarks' have in common? Answer: Not one is a Congressional earmark. They were all enacted by President Obama and his bureaucrats."[368]

I could not agree more that these are examples of unnecessary, silly, wasteful spending. That $4.9 BILLION could certainly have been spent more wisely...or not at all!

But not one of these expenditures was authorized by the Senate. They were expenditures authorized, and spent, by President Obama and his bureaucrats because the Senate did not do its constitutional duty. *They were bureaucratic earmarks!*

Why do I place the burden of blame on the Senate for these expenditures? Because the House Republicans had invoked a one-year moratorium on "earmarks," defining "earmarks" in the House rules, which I will explain later. The House Republicans not only initiated a "year-long ban" on earmarks, but encouraged both the House Democrats and the Senate to do the same. This Republican decision in the House came a day after House Democrats said they wouldn't fund special projects for defense contractors, energy firms, or private companies in general.

Democrat Congressman David Obey said, "The political reality right now is that the public has lost some confidence in this institution and one of the reasons is the past abuse of the earmark process."[369]

My reaction to Congressman Obey was that the earmark process is not what has caused the public to lose confidence. The public has lost confidence in the legislative branch when it comes to spending issues because we have created deficits in discretionary spending and we have failed to tackle the GIANTS of spending—the entitlement programs that continue to grow without restraint.

The Democrat Senator Daniel Inouye, Chair of the Senate Appropriations Committee, agreed to a two-year moratorium on earmarks, explaining, "The handwriting is clearly on the wall," Inouye said in a statement. "The president has stated unequivocally that he will veto any legislation containing earmarks."[370] Of course he would. He and his bureaucrats can still spend it. Big win for President Obama.

So there we have it. Democrats and Republicans alike in the House of Representatives and in the Senate jumped on a phony "abolish earmarks" bandwagon. So let's examine what they, the House Democrats and Republicans, are saying.

Those who advocate the end of Congressional earmarks are saying, in effect, "Let's give all authority for the function of government programs to a centralized executive branch."

While there is spending that should be refused appropriation or authorization, an across-the-board ban is wrong. Items should be defeated by Congress on the basis of their substance, not as part of a sweeping ban on all appropriations and authorizations. Let me give you examples. Improved armor for our soldiers with MRAP vehicles (Mine Resistant and Ambush Protected) and Unmanned Aerial Vehicles, such as the Predator drone, are examples of Congressional appropriations that have improved our national defense. In these cases, appropriations that some erroneously call "earmarks," saved lives.[371]

Banning or eliminating Congressional appropriations they call earmarks, simply sends the money to the Executive Branch. The expenditure of taxpayer dollars remains at the same level, only with bureaucrats and administration officials allocating the funds rather than Congress. Not one dime is saved.

The Democrats are more than happy to see Republicans abandon their authorization and appropriation responsibilities. It allows the current Democrat-run executive branch to spend, spend, spend.

I was appalled at what the House Republicans chose to do. It was flat-out wrong on several accounts. First, the resolution they passed was this: "Resolve, that it is the policy of the Republican Conference that no Member shall request a congressional earmark...as such terms are used in Clause 9 of Rule XXI of the Rules of the House for the 111th Congress." Clause 9 of Rule XXI applies to all legislation in the House of Representatives, whether it be authorization, appropriation, tax, or tariff legislation.[372]

THE REVERSAL OF A SOLEMN OATH

It was clear to me that these elected officials abrogated their responsibility, and their actions resulted in a large expenditure of money that they *could* have stopped. I see a real dereliction of duty.

Every House Republican who voted for the resolution was a Member of Congress who had taken an oath and solemnly swore, "I will support and bear true allegiance to the Constitution of the United States...so help me

God." In the wake of this Republican resolution in the House, some sought to make a case that the Republicans had just trashed their oath of office and the Constitution. And they were right. But, as I mentioned previously, the House Democrats jumped on the bandwagon and were equally guilty.

THE VALUE OF DEBATE ABOUT SPENDING

One of the main benefits of spending in the hands of legislators is that national programs can be better tailored to fit regional needs. It is the rightful role of Congress to participate in directing federal funds to the areas where the funds are most beneficial.

In 2008, Congress was faced with the task of appropriating and authorizing the spending of funds intended to pay for transportation programs, including congestion mitigation. That year, Congress distributed the millions of dollars in the program's funding through Congressional earmarks for one hundred projects in thirty-five states. The year before, when no Congressional earmarks were permitted, the Transportation Department (of the Executive Branch of government) funded projects through bureaucratic earmarks through grant competition. It didn't appear that much competition occurred. All of these bureaucratic earmarks went to only five big cities—Miami, Minneapolis-St. Paul, New York City, San Francisco, and Seattle. All of these cities, by the way, were considered to be Democrat strongholds. And that was during the Bush Administration! It doesn't really matter who sits in the White House. If Congress doesn't do its job, the Executive-Branch bureaucrats who hold jobs for a lifetime are able to wield a great deal of economic power. Their political bias cannot be challenged or stopped, unless Congress steps in *before* they have a chance to distribute funds.

A DIVERSION FROM THE BIGGER ISSUE

The real issue is the bloated government budget, mostly in the area of entitlements and other large programs. Rather than take on those spending behemoths, the media and others are turning attention on smaller amounts that are disturbing, but certainly miniscule in size compared to the sink holes of entitlement programs. And not just entitlements. How about the $700 billion TARP, nearly $890 billion stimulus, and the $2.5 trillion Obamacare? I'm embarrassed to admit that TARP was the Republicans' fault as much as the Democrats. On October 1, 2008, seventy-four senators including thirty-four Republicans gave an unelected bureaucrat $700 billion with no accountability. I call it Group Therapy. People will forget, and they did. What did they do with the TARP money? Bailouts: AIG, Chrysler, GM, and the rest of them. And the very Republicans who voted for the $700 billion TARP spent hours on the Senate Floor complaining about the bailouts that TARP funded. And sure enough, everyone forgot. But they didn't forget about "earmarks." That diversionary tactic worked beautifully.

In only 2009 and 2010, of the $3.1 trillion and $3.6 trillion budgets, Congressional earmarks make up less than half a percent. However, they seem to get 100 percent of the attention.

RAUCH GOT IT RIGHT!

Jonathan Rauch wrote an article titled "Earmarks Are a Model, Not a Menace" that appeared in the *National Journal* (March 14, 2009). Here are my favorite excerpts from that article:

> Beating up on earmarks is fun. But if you interrupt the joy long enough to take a closer look, you may discover that the case against earmarks has pretty much evaporated over the past few years. In fact, reformers seem to want to hound out of existence a system that actually works better than much of what Washington does...

> As transparency has taken over, the case against earmarks has melted away. Their budgetary impact is trivial in comparison with entitlements and other large programs. Obsessing about earmarks, indeed, has the perverse, if convenient, effect of distracting the country from its real spending problems, thus substituting indignation for discipline....

> Some earmark spending is silly, but then so is some non-earmark spending, and there is a lot more of the latter....

> And earmark spending today is, if anything, more transparent, more accountable, and more promptly disclosed than is non-earmark spending. Indeed, executive agencies could stand to emulate some of the online disclosure rules that apply to earmarks.[373]

I pretty well covered the argument succinctly in an op-ed piece I did for the *Washington Times* on December 3, 2010. I wrote:

I am used to being all alone. I was the only "no" vote in the Senate on the Everglades Restoration Act. Three years later, major publications said I was right and the other 99 were wrong. In 2002, I was alone in exposing the global-warming hysteria as a hoax. Now I have been vindicated on that issue as well. As the only conservative Republican to vote against the earmark moratorium within our conference, I find myself alone once again. But, as before, I eventually will be proved right. My opposition to the moratorium is based on my concern that Congress would be ceding its constitutional authority to the president, while failing to save a single taxpayer dime and distracting from the real issue of out-of-control deficit spending.

A politically correct ban on congressional earmarks will give President Obama even greater power and authority in the expenditure of taxpayer funds. In other words, in the case of Mr. Obama, he would have more money to pursue his liberal agenda. No wonder he was so quick to endorse a ban on congressional earmarks.

With this greater power, the Obama administration will embark on its own bureaucratic earmarks, which will result in the same type of spending that we saw from the stimulus bill, which did not contain a single congressionally directed spending item. These types of presidential earmarks will mean spending millions of tax dollars for turtle walkways, toilets in national parks, research on the mating habits of insects, and equipment to find radioactive rabbit droppings. Lobbyists already have been hitting up federal agencies at increased rates. A ban on congressional earmarks will only further increase the number of lobbyists seeking influence with the executive branch.

Congress would then be nothing more than a rubber stamp for Mr. Obama's spending requests. Transparency, accountability and the public's recourse would greatly diminish. Currently, members of Congress must make public notice of their spending requests in advance, then be held accountable by voters. However, a ban on congressional earmarks would result in the public being kept in the dark until a year or more after a presidential earmark already has been spent. At the same time, voters would be powerless to hold faceless bureaucrats accountable. What's worse is that the whole process would be pushed into Washington's darkest corners, outside the public's purview.

That is not how our Founding Fathers envisioned our government. That is why, when writing the U.S. Constitution, they gave Congress, not the executive branch, the power of the purse. Writing in the Federalist Papers, James Madison noted that Congress holds this power for the very reason that it is closer to the people. Supreme Court Justice Joseph Story noted in 1833 that if this authority were given to the president, "the executive would possess an unbounded power over the public purse of the nation; and might apply all its monied resources at his pleasure. The power to control, and direct the appropriations, constitutes a most useful and salutary check upon profusion and extravagance, as well as upon corrupt influence and public peculation." Congress should not cede this authority to the executive branch.

Minus real reforms from Congress to reduce spending, federal spending will continue to spiral out of control. Why? Because banning congressional earmarks won't save a single taxpayer dime. If an appropriations item that is directed by Congress is removed (or an attempt is made to remove the item), the money does not return to the Treasury to pay down the debt. Instead, the bottom-line expenditure amount remains the same, and the money is put into the hands of the executive branch, in this case, Mr. Obama, to spend how it sees fit. Given that the overall number and dollar amount of earmarks has decreased steadily over the past several years while the federal debt has increased by $3 trillion in just two years, an earmark ban is not the answer to our fiscal problems.

Eliminating all earmarks would have additional consequences. Vitally important earmarks, such as to provide improved armor that has saved lives for our troops in Iraq and Afghanistan, and the Predator drone program, which has been vital in the war on terrorism, would not be possible if the ban were in place. Both are examples of congressional earmarks that never would have been funded by the administration.

Let me explain it a different way. This is how it is supposed to work. The President submits his budget to Congress. The authorization committee then evaluates that budget and makes changes within the President's bottom line. In other words, the authorization committee recommends to Congress the areas where we can best provide funding for items in the President's budget. Let's take an example: I serve on the U.S. Senate Armed Services Committee which is an authorization committee staffed with defense experts from strike vehicles to missile defense. Before the ban on congressional earmarks, the President's budget contained $350.6 million for a launching system referred to as a "box of rockets" (otherwise known as the Non Line of Sight—Launching System). The Senate Armed Services Committee agreed that the launching system was good but we had a greater need for strike vehicles. So we struck the launching system and used the money to buy six new F18 E and F aircrafts. It didn't cost any more, it just redirected the money to something that would enhance our defense system in a greater way.[374]

But wait. These kinds of changes can be characterized as earmarks. However, without these kinds of changes, the President would make all spending decisions. This is not what Article 1 Section 9 of the Constitution says. The legislative branch is supposed to do the authorizing and appropriating. To be clear, many things that are proposed to be authorized and appropriated should be defeated. But we should defeat them based on the substance, not simply because they are called earmarks. I continued in my op-ed:

> Unfortunately, the years of demagoguing earmarks have distracted the American people from the real fiscal problems that face our nation. We must do something to stop runaway spending. Ironically, the authors of the ban both supported the $700 billion bailout and the $50 billion President's Plan for Emergency AIDS Relief bill, two of the largest measures of 2008. That same year, the Office of Management and Budget calculated total earmark spending at $15 billion. So, by supporting just those two measures, they obligated the government to 50 times the total of all earmarks for that year.
>
> There is a simple solution to the earmark problem that I have been advocating for more than five years. All we have to do is redefine "earmark" as spending that has not been authorized, meaning it has not been approved by the committee of jurisdiction. Then eliminate all earmarks—no exceptions. That's all. Problem solved. Then we can go after the real problems, the big stuff like the debt and the deficit.
>
> Let me repeat. The only reason I bring up the earmark issue is to demonstrate that my decade-long fight against global warming has ALSO been a simultaneous fight against bureaucratic earmarks, and the unwise spending of hundreds of millions of taxpayer dollars.[375]

I applaud Congressman Ron Paul (R-TX) in being one of the only conservative Republicans in the House to defy the demagoguery and join me. Congressman Paul and I recently issued a joint new release titled "Earmark Ban a Huge Victory for Obama" in which we said the following:

> The current ban on congressional earmarks gives the Obama administration and federal bureaucrats even greater power and authority over the expenditure of taxpayer funds. It's no surprise, then, that Mr. Obama was quick to endorse the ban on congressional earmarks—it grants his administration more power to pursue its agenda by exercising a power that properly resides with Congress. The president even endorsed the ban with his veto threat during this year's State of the Union Address and in subsequent speeches.

> With this greater power, the Obama administration (and future administrations) will embark on its own bureaucratic earmarks through a process that will mean less transparency and little accountability for the American people. The infamous $787 billion Stimulus bill, which both of us vigorously opposed and voted against, did not contain a single Congressional earmark. Instead, millions in bureaucratic earmarks were spent for programs to determine the affects of intoxication on mice, the protection of insects from other insects, the mating habits of bugs and rodents, and walkways for turtles.

> We were the only two conservatives to be outspoken against the current ban, because it was the right thing to do—politically unpopular, but the right thing. With the current earmark ban in place, Congress becomes nothing more than a rubber stamp for President Obama's spending request, and our opposition to the ban is based on the fact that this is not how our Founding Fathers envisioned our government. When writing the U.S. Constitution, they gave Congress, not the executive branch, the power of the purse.

> According to James Madison's view outlined in the *Federalist Papers*, Congress holds this power for the very reason that it is closer to the people. Supreme Court Justice Joseph Story noted in 1833 that if this authority were given to the president, "the executive would posses an unbounded power over the public purse of the nation; and might apply all its monied resources at his pleasure."

> The Constitutional power has been ceded to the president.

What's worse is the fact that banning Congressional earmarks won't save a single taxpayer dime. Instead of saving taxpayer money, the spending authority is shifted from Congress to the Executive Branch. In this case, the money is put into the hands of President Obama, to spend how he sees fit.

Unfortunately, the years of demagoguing earmarks have distracted the American people from the real fiscal problems that face our nation.

Is it any wonder that Obama supported the big push behind the ban on congressional earmarks? With Congress out of the way, he is now the king of earmarks.[376]

NOW THE GOOD NEWS

There is some good news. Senator John McCain and other conservatives have joined me in a solution to the earmark issue. We have introduced legislation to redefine "earmarks" as "an appropriation that has not been authorized." That should solve the Congressional earmark problem.[377]

As Senator McCain stated on the floor, "Some of those earmarks are worthy. If they are worthy then they should be authorized."[378] He also said, "You've got to get the definition of an earmark: that is, an unauthorized appropriation."

And as Senator Coburn (R-OK) said, "It is not wrong to want to help your state. It is not wrong to go through an authorizing process where your colleagues actually see it."[379] He and I believe if something is really bad you have two chances to kill it: in authorization and appropriations.

So, I repeat, the only reason I bring "earmarks" into this book, is because bureaucratic Obama earmarks—unelected bureaucrats putting grant money into liberal causes, brainwashing the public—has been a major obstacle I have had to overcome in combating the hoax.

I will never forget when, years ago, my granddaughter came home from school and asked, "Why is it you don't understand global warming?" She had been brainwashed by a grant. We traced what she had been taught all the way from the EPA to our public school system in Oklahoma—the false information was the result of a grant designed to brainwash our kids, using a bureaucratic earmark.

One of the great frustrations is that it will take years for conservatives and some talk show hosts to appreciate this chapter. If you ask the vast majority of talk show hosts, or even newly elected and many long-term members of Congress to define earmarks, they cannot do it. But, even though they may know better, many politicians have continued to mischaracterize earmarks, often out of a desire for personal political gain.

Webster defines a demagogue as "a leader who makes use of popular prejudices and false claims and promises in order to gain power."[380] Demagoguery is popular and sends approval numbers soaring.

For the first seven years of my ten-year battle to stop the global warming cap and trade bills no one would join me because they were convinced people had already made up their minds. They said taking it on was politically stupid, and they were right. I went through seven years of misery. But we won. And we are continuing to win. And we are determined to win until there are no more global-warming lies, cap-and-tax tricks, and phony-baloney Obama bureaucratic earmarks to battle.

YOU DECIDE WHO RULES

In the end, this issue of an obvious executive-branch power grab is an issue the voters will have to decide.

Do you want *elected* officials to pass laws? These officials are *your* representatives, sent to Washington, D.C., to do *your* business.

Or do you want *appointed* bureaucrats to establish regulations over which you have very little recourse?

Again, we need to keep the authorization and appropriation process in the hands of Congress where the Constitution placed it and go after bureaucratic earmarks. It is a fact bureaucratic earmarks have been just as much a part of the global-warming "lie machine" as Al Gore's movie and they need to be eradicated the same as Al Gore's science fiction movie has been refuted.

APPENDIX A

WHAT'S IN IT FOR THE UNITED NATIONS?

SUSTAINABLE DEVELOPMENT AND THE QUEST FOR AUTONOMY...WHERE GLOBAL WARMING BEGAN

The United Nations was founded after World War II to replace the League of Nations. Its expressed purpose at that time, in 1945, was to help nations work together and talk to avoid wars and promote social progress.[381] The *stated* aims of the UN have remained fairly stable over the decades, at least officially. Article 1 of the UN Charter states that the UN is to, among other things, "maintain international peace and security[382]...develop friendly relations among nations[383]... achieve international cooperation in solving international problems of an economic, social, cultural, or humanitarian character,"[384] and to serve as "a center for harmonizing the actions of nations."[385] Nations hoped that by encouraging cooperation on these issues, countries would be moved toward peaceful relationships instead of warring ones.

I believe that many globalist elites have worked within the United Nations to expand its responsibility to an alarming degree. Now, instead of facilitating international cooperation, I believe the UN's primary institutional goal—in practice, if not in word—is to actively build a global utopia. The UN believes that it can—with enough power and influence—determine what is best for the world by reaching agreements by majority agreement, or better yet—consensus—among all of the member states participating at the United Nations.

Each country represented at the UN has an equal voice.[386] No nation—regardless of their population, landmass, or global influence—has any more power at the United Nations than another. Each country has one vote.[387]

When the United Nations works to pass resolutions, it is unwavering in its desire to do so by consensus. I think that the UN wants all nations to agree because many elites within the UN believe that where you find agreement, there you will find peace. If the sum of all nations can agree how to tackle common problems, then global peace should materialize with relative ease. The UN proclaims that its adamancy toward reaching agreement by the consensus of all nations gives it "moral authority" over all other institutions. This, the UN describes, is one of its "best properties."[388]

Over the past several decades, the United Nations has turned much of its attention to the crafting of solutions for two problems I believe it has designated as high priorities. It has worked tirelessly to provide a consensus-based solution for them. The environment is one. Development is the other.

According to the bureaucratic elites at the UN, the environment is on the brink of disaster. They believe that anywhere you go, you will find evidence of a global population paying little attention to the impact their activities have on the environment, and caring even less about the long-term consequences of those actions.

Concurrently, these elites see billions of people around the world living in absolute poverty. The global poor—heavily concentrated in the least developed countries—often live on just dollars a day and have extreme difficulty living even subsistence lifestyles. All the while, they watch those in developed nations enjoy the prosperity their wealth provides, but see them paying little attention to the social dilemma inherent in this inequality.

On a fundamental level, the United Nations elites see three core spheres of humanity: the society, the economy, and the environment, and they see major problems in all three. They believe society has little regard for the inequality across people groups around the world. To them, the global economic structure appears to be skewed in favor of those already possessing and consuming most of the world's resources—leaving little left over for the poor. The environment has been disregarded by many in favor of faster economic growth—to the peril of future generations.

The UN elites believe the condition of these three spheres is unsustainable; they are out of balance; causing inequality; and risking serious, irreversible damage to the global community. As long as they remain out of sync, they believe that establishing global utopia will be impossible. Conflicts—armed or otherwise—will undoubtedly sprout from this unsustainable trajectory, underscoring the need to develop a workable solution to reverse this alarming trend.[389]

Sustainable development is the guiding philosophy that the UN elites have constructed to solve the structural problems they have identified, and they seek to use this philosophy to bring complete harmony to each of humanity's three spheres.

Doing this will demand fundamental and comprehensive change to a number of international frameworks, and fully implementing the philosophy could put the UN on a dangerous path towards autonomy. This is not something we can allow to happen.

SUSTAINABLE DEVELOPMENT & SOCIETAL CHANGES

The United Nations' elite believe that the core of the environmental problem facing the world is one of societal outlook. Al Gore articulated this well when he wrote that the "the twentieth century has not been kind to the constant human striving for a sense of purpose in life. Two world wars, the Holocaust, the invention of nuclear weapons, and now the global environmental crisis have led many of us to wonder if survival—much less enlightened, joyous, and hopeful living—is possible."[390] He believes—and the UN elites agree—that this questioning about the future has caused many to recklessly seek material gain without any regard for the potential consequences to the environment or society.

Solving this problem at the philosophical level will require the restoration of faith in the belief that we *do* have a future that is worth preparing for. Gore believes this is "essential to [restoring] the balance now missing in our relationship to the earth."[391]

As a lawmaker, I am keenly aware of the challenges facing anyone who wants to change another person's behavior. Solving a philosophical problem rooted in faith will prove impossible for an institution like the UN. But the UN elites have worked hard to outline the changes they believe need to happen within the different spheres of society to bring everything back into balance.

Widespread modern concern for the environment began in the 1960s and 1970s. In 1962, Rachel Carson wrote her landmark book *Silent Spring*, questioning the actual benefits of DDT, a widely used pesticide at the time. The book led to broader public questioning of using pesticides and chemicals in everyday life.

In 1968, Paul Ehrlich published another influential book, *Population Bomb*. In it, he predicted that mass global starvation would occur as a result of overpopulation. He did not believe that global food production would expand as rapidly as the global population, a theory he based on the idea that the earth has a predictable carrying capacity that can sustain only a certain number of people at a middle-class standard of living. To address the problem, he recommended the immediate implementation of population control policies.[392]

Carson's theory questioned whether or not all human advancement is truly beneficial. Ehrlich's work sparked skepticism that the earth has the ability to sustain widespread global industrialization. Their common theme is that they coincided with and heavily influenced a rising concern about humanity's overall impact on the environment.

Al Gore is one leader who was impacted by the work of Rachel Carson. He wrote that dinner table conversations during his childhood about her book "made an impression" on him because they made him "think about threats to the environment that are much more serious than washed-out gullies—but much harder to see."[393] This revelation would make him sympathetic to emerging global environmental problems like global warming and the ozone hole, which predicted impending disaster because of the release by humans of seemingly harmless chemicals and compounds.

Believing that the earth could be heading down a path toward environmental catastrophe caused the elites within the UN to push for action.

Their involvement formally began with the planning of the 1972 Conference on the Human Environment, which would ultimately serve as the birthplace of the sustainable development philosophy.

Sustainable development aims to solve the problems caused by environmental degradation by organizing society in a way that preserves the present status of the environment. The UN elites believe that doing this will preserve an environment suitable for future generations to provide for their own needs. The UN elites further believe that enshrining an environment that is sustainable demands that the natural systems that support life on earth be preserved as development occurs. These systems include "the atmosphere, the waters, the soils, and the living beings."[394]

The elites contend that limiting the rate of depletion of nonrenewable resources is an important component of sustainable development because of the predictable carrying capacity theory. According to the theory, future generations will be unable to meet their development needs if today's needed resources have been exhausted. Limiting the depletion of nonrenewables to the lowest level necessary to sustain society is the goal, as it will result in the longest enjoyment of those resources. Because these resources, like oil and coal, will eventually be depleted, sustainable development has an overwhelming bias toward the immediate adoption of renewable resources and recycling patterns, no matter the cost. The UN elites believe that the adoption of consumption and production patterns respectful of these trends will ensure the availability of resources—and therefore equal opportunity—for future generations.

The predictable carrying capacity theory also doubts the ability of the earth's natural systems to support complete industrialization of the world's population. In other words, it fears that the unintended consequences of industrialization, mainly global warming, will push the delicate balance of the earth out of whack and into disaster.

This presents an interesting ethical dilemma. If the earth cannot handle the rising living standards of the global poor, should those who are wealthy be allowed to maintain lavish lifestyles?

The UN elites believe that such *intragenerational* inequality is unfair. As such, the UN believes that accomplishing sustainable development requires a standardization of living standards that is sensitive to the earth's predicted carrying capacity. The elites are fully aware that this will demand a reduction in the living standards of the populations in developed countries. *Our Common Future*, a key UN sustainable development document that is heralded by the elites, states that "living standards that go beyond the basic minimum are sustainable only if consumption standards everywhere have regard for long-term sustainability…sustainable development requires the promotion of values that encourage consumption standards that are within the bounds of the ecological possible and to which all can reasonably aspire."[395]

In other words, the UN elites believe that they need the power to determine—and enforce—an appropriate global standard of living that is sensitive to the earth's predictable carrying capacity. In doing this, the UN believes it will be able to ensure that no one individual's development potential is cut short by the lavish lifestyles of others.

SUSTAINABLE DEVELOPMENT & CHANGES TO SOCIETY'S ECONOMIC STRUCTURE

Developing nations have not always gone along with the UN's plans to limit their ability to pursue development. This is not surprising. As the home to most of the global poor, developing nations should be looking for every development avenue possible to improve the living standards of the people.

Initial discussions leading up to the UN's first environmental confer-ence, the 1972 Conference on the Human Environment, were mainly focused on the problems of the developed world. And because economic growth was seen by many as a key cause of environmental degradation, some of the solutions being considered included the idea of slowing eco-nomic growth. Developing nations feared that the effort to limit environ-mental damage would yield policy changes that curtailed their ability to develop, which is what they needed to harness to climb out of poverty. If the fragility of the environment was the only concern of the conference, then they wanted no part of it, so they threatened to boycott the event.[396]

The potential balk by developing nations was important to the UN elites for one reason: the UN's desire to craft agreements by majority. Under the institution's one-nation, one-vote system, developing nations have extraordinary power to influence and frame whatever discussion takes place at the UN. This is because they make up a large majority of the total number of member states. Without the support of developing nations, a UN resolution cannot even obtain a simple majority, let alone consensus.

The concentration of this power occurred in 1964 when seventy-seven nations banded together to create the Group of 77 (G-77). This group works at the UN to advance the goals and agenda of the developing world. Since then, its membership has climbed to a total of 131, a significant majority in the UN's General Assembly.[397]

A threat by the G-77 to avoid the 1972 Conference on the Human Environment, therefore, had to be taken seriously. Without their support, there would be no way to reach a consensus—or even a majority—and the UN's effort to address the global environmental problem would be significantly weakened.

To garner their support, conference administrators, led by entrepre-neur-turned-environmentalist Maurice Strong, agreed to amend the draft conference framework so that it would be more inclusive of the developing world's concerns. To figure out how to do this, he decided to hold an informal meeting in Founex, Switzerland, in 1971, where he and the conference participants would answer any questions and put to rest any concerns that there may be a conflict of interest between environmental protection and economic growth/development.[398]

Strong was able to allay developing nations' fears by incorporating their development ideals into the broader fight to save the environment. He did this by expanding the definition of the word "environment." The Founex Panel ultimately resulted in the environment no longer being considered "simply the biophysical sphere," but instead it would also include the "socio-economic structures." The Founex Panel stated that the bio-environment and society "[form] an interdependent and inextricable web."[399] No longer would the debate on the environment be "concerned only with pollution and conservation, nor was the damage to the environment attributable solely to the process of development. In many cases, the damage was due to the very same socio-economic forces and causes that were at the root of poverty, underdevelopment, and inequality, and could be overcome only through the process of development, economic growth, and social change."[400]

The tangible result of Founex was the modern framework of sustainable development. It shifted the entire "environmental" debate to become more centrally focused on the development of impoverished nations.

Throughout the history of the UN, the G-77 has worked to announce its frustrations with the global economic power structure. In 1974, developing nations worked together to pass a UN resolution declaring the need to establish a New International Economic Order.[401] In short, the resolution blames colonialism and the western-style capitalistic system as the root cause of the impoverished state of their own countries. They point to the fact that the vast majority of the world's resources are consumed by a relatively small group: the people living in industrialized countries. Developing nations believe this gives them little ability to procure the resources necessary for development. This was inherent in the Cocoyoc Declaration, which was made the same year, which states that "pre-emption by the rich of a disproportionate share of key resources conflicts directly with the longer-term interests of the poor by impairing their ultimate access to resources necessary to their development and by increasing their cost."[402] In other words, they believe the economic development being pursued by developed nations pushes the price of raw materials up to a level that is unaffordable to those in developing nations. They believe this limits their potential to develop.

They further state that the developed nations use their enormous power in the global economic system to construct trade and other international agreements to be in their favor, again at the expense of developing nations' development.

Sustainable development aims to directly combat these challenges to development that it believes are fostering widespread global inequality by encouraging the restructuring of the global economic landscape in favor of developing nations. It also aims to provide more ready and affordable access to the resources necessary to accomplish development.

Establishing the right to development is one of the key ways that sustainable development does this. This right demands that all nations and people be guaranteed the benefits of development. To the extent that developing nations cannot provide the benefits of development for their people, sustainable development implies that the global community should provide it for them.

Because affordable access to raw and other materials is required to competitively construct goods for trade and consumption, sustainable development demands that these resources be provided to developing nations affordably. Access to technology is another resource that must be secured affordably to take advantage of the maximum benefits of development. To the extent that individuals in developing nations cannot afford the most advanced technology available, it should be made available to them for free, or at a price they can afford.

Because so much of the world's resources are controlled and consumed by developed nations, they maintain a dominant hand in international economics. The UN bureaucracy believes developed nations like the United States use this influence to unfairly construct trade agreements and multinational financial institutions, like the IMF, to be overwhelmingly biased toward their own goals. The UN believes that this causes significant costs and barriers to development for developing nations. Consequently, sustainable development demands the realignment of these agreements and institutions to be overwhelmingly in favor of developing nations so that they can more easily enjoy the benefits of development.

In short, sustainable development demands the removal of global wealth and income inequality by completely shifting the global economic landscape in the favor of developing nations. In many ways, doing this allows the environmental leaders at the UN to simultaneously accomplish its societal goals detailed earlier.

The regulation of raw materials and their guaranteed distribution in accordance with sustainable development needs would require control over natural resources supplies. To the extent that demand cannot be controlled, supply must be regulated.

This control over all resources would ultimately allow the UN to reduce the developed nations' living standards so that they would be in line with the UN's prediction of the world's carrying capacity. This would protect the environment. Simultaneously, the UN could redirect the extracted materials toward the needs of the developing world so that their standards of living can rise to that set by the earth's predicted carrying capacity. At its most basic level, this regime aims to eliminate global inequality through socialism.

Once all people have reached this "sustainable" standard of living, the UN would then be able to maintain sustainability by directing economic growth and technological change on a sustainable path, allowing standards of living to rise only as technological advancement and the discovery of new resources allow.

Ultimately, the environmental corps at the UN believes that pursuing sustainable development will reduce conflict and promote peace by restoring equity to the relationships between the earth and between fellow man. With sustainable development, there will be no need for conflict, because the global conditions that so often lead to conflict would be permanently suspended in harmony.

IMPLEMENTATION & AUTONOMOUS POWER

Considering that the actual implementation of sustainable development is daunting, it would require nothing short of total control over the earth's resources. There is simply no other way to guarantee environmental security, global equality in living standards, and equal potential for development across the world.

As discussed before, the UN elites believe that the inherent nature of the environment's threat is one that is global and deep within society. Because the problem is global, the consequences of unsustainable development are also global. Enacting sustainable development as a corrective instrument demands a comprehensive shift in the way decisions are made within the international community so they are responsive to the global problems we face.

Convincing the world to surrender sovereignty to an international body will not be easy, and the environmental lobby knows this. History shows that the global community rarely cooperates in a united way in the absence of an impending catastrophe. Al Gore once observed that motivating nations to act collectively "has usually been secured only with the emergence of a life-or-death threat to the existence of society itself."[403] I believe that this understanding has led the UN elites to use alarmism to scare the public into action with the claim that catastrophe will be imminent without immediate implementation of sustainable development. Without this life-or-death threat of unsustainable development purported by the United Nations, there would be no need for their autonomous control over global economic resources.

This is a key reason why the global warming issue fits so perfectly within the plans of those at the UN who want to use the institution to construct utopia. Without a tangible, impending disaster, there is no catalyst to act. Without a catalyst, the UN's quest to restore equity to relationships and establish utopia would die.

The formal development of the sustainable development philosophy occurred over a period of about thirty years, between the late 1960s and the early 1990s. Since then, the sustainable development philosophy has infected nearly every United Nations initiative, and the body has used its indefinable goal of "societal equity" to bring forth policy problems of all sorts so that the UN can prescribe the solution.

A few examples:

In 1995, the UN held a World Summit for Social Development. The documents released as a result of the summit directly reference the establishment of sustainable development as the overriding goal of the summit's policy proposals. Among its recommendations is the regulation of multinational corporations for the purposes of making economic growth more conducive to social development in developing countries. It also calls into question the accumulation of wealth, and proposes that it be taxed for the purposes of improving stability in financial markets, as if it caused instability in the first place.[404]

In 1996, the UN held its Second United Nations Summit on Human

Settlements. There, the Summit reaffirmed the UN's commitment to making the provision of "adequate shelter" a basic human right. In doing so, it elevates this issue to the international level, which is unnecessary. To ensure this right is protected, it states that it is the responsibility of governments to "enable people to obtain shelter and to protect and improve dwellings." While this certainly does not sound bad, it suggests that the government should be in charge of assigning housing to people of low income. The conference leader also suggested that each nation cut its defense budget by 5 percent to address the housing needs in an affordable way.[405]

In 2002, the UN produced a document called "A World Fit for Children" that reaffirms the rights established for children at the Convention on the Rights of the Child. It made clear that the right to sexual health privacy is a necessary element of sustainable development. Without this right enacted, unsustainable development will ensue, which in their opinion is a bad thing.[406]

There is no shortage of alarming changes demanded by the UN in the name of accomplishing sustainable development, and since launching sustainable development, the body has stepped well beyond the original intuitions of what the philosophy should accomplish. It has, in practice, become the catch-all method by which the UN allows itself to elevate any problem it identifies within society to the international level. The mentality that it can solve all the world's problems and bring about utopia is centered in the philosophy that governing by the consensus of nations is the surest way to avoid conflict.

WHY WE SHOULDN'T PURSUE SUSTAINABLE DEVELOPMENT AT THE UN

Surrendering sovereignty to an international body like the United Nations so heavily influenced by environmental elites pursuing a utopian agenda of sustainable development is a dangerous idea. From the United States' perspective, the sustainable development agenda provides no benefit.

Its goals to restore relationships between the earth and humanity, and its attempt to restore equity between and among generations are far-fetched and are not grounded in reality. The societal changes demanded to improve environmental wellness are all based around the idea that the earth's carrying capacity is predictable. I believe that the earth has a carrying capacity to some degree, but I also believe that our knowledge of the future is severely limited. Who would have thought two hundred years ago, that we would be able to feed our entire country with less than two percent of the entire workforce dedicated to agriculture?[407] This has been made possible because of technological and other advances that have improved our productivity. Why should we expect this trend to stop?

The Industrial Revolution and the Information Age have made technological change occur at incredible speeds. Consequently, our demand for resources changes every minute. Further limiting already limited non-renewable resources in the name of sustainable development will only push the price of those materials up as we continue to use them. This will make products—and development—more expensive, which is the exact opposite of what many at the UN want sustainable development to do.

The economic changes demanded by the implementation of sustainable development are equally befuddling. Giving developing nations a right to development, per se where the international community pitches in to provide for a country's needs without regard to the decisions made by those developing nations, is a bad idea. I believe that it is in the best interest of the United States to help developing nations prosper. I also believe it is our moral obligation. Assistance, however, should never come without strings attached.

In America, we know that our prosperity is a direct result of the government founded by our forefathers. They understood the value of a stable government with limited, separated powers, and they insisted that the freedom of the individual be protected. These are a few of the many characteristics of our nation that have provided us with significant material blessing.

Similarly, many nations have pursued highly damaging economic policies—like totalitarianism and communism—and they have reaped what they have sown.

Promoting a right to development, where the international community pitches in and attempts to direct and guarantee prosperity, overlooks this core truth. Nations must have the ability to choose whatever development path they want, *and they must also live with the consequences of their decisions.*

The right to development, by undermining the need for accountability, runs the risk of propping up bad leaders in foreign countries. If this development aid is given to nations whose leaders are awful, such that the benefits of aid can be more easily felt by a nation's people, then the nation's people will be less outraged at the awful decisions made by their own government. This will dull their appetite to demand leadership changes, which are often necessary in bad situations like these.

Instead of encouraging a right to development, the United States and other countries should look to help developing nations improve their institutional rules, frameworks, and governing philosophies to be friendlier toward business, tougher on crime, and stable for the long run. Implementing these changes will go much further to help developing nations take advantage of development than hand outs and bailouts ever could.

Similarly, the assertion that the wealth of developed nations has caused the developing nations to be subjected to poverty is wrong. This argument has been made since at least the mid 1970s, and since that time it has been refuted. P.T. Bauer, a renowned economist, wrote in 1977 that developing nations with the most open relationships with developed countries saw the most economic growth. Nations that were the most isolated—those with the fewest connections to the West—experienced worse economic growth compared to their peers.[408] For a modern example, we need to go no further than the Korean Peninsula. North Korea, with a communistic/totalitarian state, has isolated itself from the world. It is one of the poorest nations on earth, and its people suffer from severe poverty. South Korea, on the other hand, is a democracy that values freedom. It has opened itself up to relationships with the West, and it is now one of the most advanced economies in the world. This fact single handedly disarms the argument that the success of developed nations is preventing developing nations from enjoying the benefits of growth and development.

Economic growth occurs because individuals are free to make decisions of personal benefit in an open, safe, and reliable society. This means that there is no need for a centrally controlled system of raw materials. Such a system might try to shuffle resources from developed nations to developing nations, but in the end, it will yield an economic system that is irrational and not properly sensitive to the changing needs and desires of the seven billion people on the planet.

CONCLUSION

The desire of many to work through the UN to establish utopia under the ideals of sustainable development is real. The desire is also dangerous. Past attempts to build utopias have failed miserably, and they have often resulted in mass murder and genocide. Promoting an agenda that demands the centralization of power into the hands of a single institution is a recipe for disaster. We must remain vigilant and aware of what the UN is doing to accomplish these goals. As you've learned from reading this book, the clearest and most direct way they're trying to do this is through the global warming agenda and the pursuit of a Kyoto-like treaty. A treaty of that magnitude would require a true shift of power from sovereign nations like our own to the United Nations so that they can determine how to distribute our wealth and resources around the world—in a way that meets their definition of "fair." We must remember to look beyond the headlines and press releases and do the hard work required to dig deeply into UN documents to understand their true intentions and motivations. By doing this, we will be able to more effectively prevent our nation's sovereignty from being eroded by the goals of the super-liberals at the United Nations. To that end, we must demand that the UN remain open and transparent about all that it does.

Doing anything less will result in a severe threat to our own freedom and prosperity.

APPENDIX B

EXCERPTS FROM MICHAEL CRICHTON'S NOVEL: *STATE OF FEAR*

*All excerpts are copyrighted and printed with permission from Michael Crichton, State of Fear, Harper Collins, 2004.

EXCERPT #1:
The Political Manipulation of the Science.

Dr. Crichton clearly understood that the science was being manipulated from the very beginning.

"'IPCC? What last minute changes?'

"The UN formed the Intergovernmental Panel on Climate Change in the late 1980s. That's the IPCC... a huge group of bureaucrats, and scientists under the thumb of bureaucrats. The idea was that since this was a global problem, the UN would track climate research and issue reports every few years. The first assessment report in 1990 said it would be very difficult to detect a human influence on climate, although everybody was concerned that one might exist. But the 1995 report announced with conviction that there was now a 'discernible human influence' on climate."

"'...a discernible human influence' was written into the 1995 summary report after the scientists themselves had gone home. Originally, the document said scientists couldn't detect a human influence on climate for sure, and they didn't know when they would. They said explicitly, 'we don't know.' That statement was deleted and replaced with a new statement that a discernible human influence did indeed exist. It was a major change."

"... Changing the document caused a stir among scientists at the time, with opponents and defendants of the change coming forward. If you read their claims and counter-claims, you can't be sure who's telling the truth. But this is the Internet age. You can find the original documents and the list of changes online and decide for yourself. A review of the actual text changes makes it crystal clear that the IPCC is a political organization, not a scientific one."

"...When Hansen announced in the summer of 1988 that global warming was here, he predicted temperatures would increase .35 degrees Celsius over the next ten years."

"...The actual increase was .11 degrees."

"And ten years after his testimony, he said that the forces that govern climate change are so poorly understood that long-term prediction is impossible."

"He said, 'The forces that drive long-term climate change are not known with an accuracy sufficient to define future climate change."

EXCERPT #2:

The Shift from Global Warming to Climate Change.

Dr. Crichton clearly saw the shift in terminology, for political purposes. Two of the characters in his novel, Drake and Henley, devise a way to keep money flowing as the issue of global warming loses credibility.

"I hate global warming," Drake said ...

"...You can't raise a dime with it, especially in winter. Every time it snows people forget all about global warming. Or else they decide some warming might be a good thing after all. They're trudging through the snow, hoping for a little global warming. It's not like pollution, John. Pollution worked. It still works. Pollution scares... people. You tell 'em they'll get cancer, and the money rolls in. But nobody is scared of a little warming. Especially if it won't happen for a hundred years."

"...Species extinction from global warming... They've heard that most of the species that will become extinct are insects. You can't raise money on insect extinctions, John. Exotic diseases from global warming—nobody cares. Hasn't happened. We ran that huge campaign last year connecting global warming to the Ebola and Hanta viruses. Nobody went for it. Sea-level rise from global warming—we all know where that'll end up. The Vanutu lawsuit is a...disaster. Everybody'll assume the sea level isn't rising anywhere. And that Scandinavian guy, that sea level expert. He's becoming a pest. He's even attacking the IPCC for incompetence."

"...Let me explain how you are going to solve your problem, Nicholas. The solution is simple. You have already told me that global warming is unsatisfactory because whenever there is a cold snap, people forget about it."

"So what you need,"... "is to structure the information so that whatever kind of weather occurs, it always confirms your message. That's the virtue of shifting the focus to abrupt climate change. It enables you to use everything that happens. There will always be floods, and freezing storms, and cyclones, and hurricanes. These events will always get headlines and airtime. And in every instance, you can claim it as an example of abrupt climate change caused by global warming. So the message gets reinforced. The urgency is increased."

"... It's not logical to say that freezing weather is caused by global warming."

"What's logic got to do with it?... All we need is for the media to report it. ... The US murder rate is as low as it was in the early 1970s, but Americans are more frightened than ever, because so much more airtime is devoted to crime, they naturally assume there is more in real life too."

" ...Think about what I am saying... A twelve-year trend, and they still don't believe it. There is no greater proof that all reality is media reality."

"... it'll be even easier to sell abrupt climate change in Europe than in the US. You just do it out of Brussels. Because bureaucrats get it, ... They'll see the advantages of this shift in emphasis."

EXCERPT #3:

The Press-Driven Agenda.

The plain and simple formula of the media business is that money is made when ratings are enhanced, and one of the sure-fire ways of boosting ratings is to introduce a strong fear factor. Dr. Crichton captures this sense of self-importance on the part of media personnel as few other authors ever have. The scene below involves three media skeptics.

Then the anchors came back onscreen, and one of the men said, "Flood advisories remain in effect, even though it is unseasonable for this time of year."

"Looks like the weather's changing," the anchorwoman said, tossing her hair.

"Yes, Marla, there is no question the weather is changing. And here, with that story, is our own Johnny Rivera."

They cut to a younger man, apparently the weatherman. "Thanks, Terry. Hi, everybody. If you're a longtime resident of the Grand Canyon State, you've probably noticed that our weather is changing, and scientists have confirmed that what's behind it is our old culprit, global warming. Today's flash flood is just one example of the trouble ahead—more extreme weather conditions, like floods and tornadoes and droughts—all as a result of global warming."

Sanjong nudged Evans, and handed him a sheet of paper. It was a printout of a press release from the NERF website. Sanjong pointed to the text: "...scientists agree there will be trouble ahead: more extreme weather events, like floods and tornadoes and drought, all as a result of global warming."

Evans said, "This guy's just reading a press release?"

"That's how they do it, these days," Kenner said. "They don't even bother to change a phrase here and there. They just read the copy outright. And of course, what he's saying is not true."

"Then what's causing the increase in extreme weather around the world?" Evans said.

"There is no increase in extreme weather."

"That's been studied?"

"Repeatedly. The studies show no increase in extreme weather events over the past century. Or in the last fifteen years. And the GCMs don't predict more extreme weather. If anything, global warming theory predicts less extreme weather."

Onscreen, the weatherman was saying, "it is becoming so bad, that the latest news is—get this—glaciers on Greenland are melting away and will soon vanish entirely. Those glaciers are three miles thick, folks. That's a lotta ice. A new study estimate sea levels will rise twenty feet or more. So sell that beach property now."

Evans said, 'What about that one? It was on the news in LA yesterday."

"I wouldn't call it news," Kenner said. "Scientists at Reading ran computer simulations that suggested that Greenland might lose its ice pack in the next thousand years."

"Thousand years?" Evans said.

"Might."

Evans pointed to the television. "He didn't say it could happen a thousand years from now."

"Imagine that," Kenner said. "He left that one out."

"But you said it isn't news..."

"You tell me," Kenner said, "Do you spend much time worrying about what might happen a thousand years from now?"

"No."

"Think anybody should?"

"No."

"There you are."

EXCERPT #4:

The Rising Role of Hollywood.

In the novel, Dr. Crichton's character Ted Bradley is portrayed in a role that accurately demonstrates how Hollywood actors rejoice in letting themselves be used to influence public opinion.

The forest floor was dark and cool. Shafts of sunlight filtered down from the magnificent trees rising all around them. The air smelled of pine. The ground was soft underfoot.

It was a pleasant spot, with sunlight dappling the forest floor, but even so the television cameras had to turn on their lights to film the third-grade schoolchildren who sat in concentric circles around the famous actor and activist Ted Bradley. Bradley was wearing a black T-shirt that set off his makeup and his dark good looks.

"These glorious trees are your birthright," he said, gesturing all around him. "They have been standing here for centuries. Long before you were born, before your parents or your grandparents or your great-grandparents were born. Some of them, before Columbus came to America! Before the Indians came! Before anything! These trees are the oldest living things on the planet; they are the guardians of the Earth; they are wise; and they have a message for us: Leave the planet alone. Don't mess with it, or with us. And we must listen to them."

The kids stared open-mouthed, transfixed. The cameras were trained on Bradley.

"But now these magnificent trees—having survived the threat of fire, the threat of logging, the threat of soil erosion, the threat of acid rain—now face their greatest threat ever. Global warming. You kids know what global warming is, don't you?"

Hands went up all around the circle. "I know, I know!"

"I'm glad you do," Bradley said, gesturing for the kids to put their hands down. The only person talking today would be Ted Bradley. "But you may not know that global warming is going to cause a very sudden change in our climate. Maybe just a few months or years, and it will suddenly be much hotter or much colder. And there will be hordes of insects and diseases that will take down these wonderful trees."

"What kind of insects?" one kid asked.

"Bad ones," Bradley said. "The ones that eat trees, that worm inside them and chew them up." He wiggled his hands, suggesting the worming in progress.

"It would take an insect a long time to eat a whole tree," a girl offered.

"No, it wouldn't!" Bradley said. "That's the trouble. Because global warming means lots and lots of insects will come—a plague of insects—and they'll eat the trees fast!"

Standing to one side, Jennifer leaned close to Evans. "Do you believe this...?"

Evans yawned. He had slept on the flight up, and had dozed off again in the ride from the airport to this grove in Sequoia National Park. He felt groggy now, looking at Bradley. Groggy and bored.

By now the kids were fidgeting, and Bradley turned squarely to the cameras. He spoke with the easy authority he had mastered while playing the president for so many years on television. "The threat of abrupt climate change," he said, "is so devastating for mankind, and for all life on this planet, that conferences are being convened all around the world to deal with it. There is a conference in Los Angeles starting tomorrow, where scientists will discuss what we can do to mitigate this terrible threat. But if we do nothing, catastrophe looms. And these mighty, magnificent trees will be a memory, a postcard from the past, a snapshot of man's inhumanity to the natural world. We're responsible for catastrophic climate change. And only we can stop it."

He finished, with a slight turn to favor his good side, and a piercing stare, from his baby blues, right into the lens.

... "Twenty thousand years ago, the Ice Age glaciers receded from California, gouging out Yosemite Valley and other beauty spots as they left. As the ice walls withdrew, they left behind a gunky, damp plain with lots of lakes fed by the melting glaciers, but no vegetation at all. It was basically wet sand.

"After a few thousand years, the land dried as the glaciers continued to move farther north. This region of California became arctic tundra, with tall grasses supporting little animals, like mice and squirrels. Human beings had arrived here by then, hunting the small animals and setting fires."

"Okay so far?" Jennifer said. "No primeval forests yet."

"I'm listening," Ted growled. He was clearly trying to control his temper.

She continued. "At first, arctic grasses and shrubs were the only plants that could take hold in the barren glacial soil. But when they died they decomposed, and over thousands of years a layer of topsoil built up. And that initiated a sequence of plant colonization that was basically the same everywhere in post-glacial North America.

"First, lodgepole pine comes in. That's around fourteen thousand years ago. Later it's joined by spruce, hemlock, and alder—trees that are hardy but can't be first. These trees constitute the real 'primary' forest, and they dominated this landscape for the next four thousand years. Then the climate changed. It got much warmer, and all the glaciers in California melted. There were no glaciers at all in California back then. It was warm and dry, there were lots of fires, and the primary forest burned. It was replaced by a plains-type vegetation of oak trees and prairie herbs. And a few Douglas fir trees, but not many, because the climate was too dry for fir trees.

"Then, around six thousand years ago, the climate changed again. It became wetter, and the Douglas fir, hemlock, and cedar moved in and took over the land, creating the great closed-canopy forests that you see now. But someone might refer to these fir trees as a pest plant—an oversized weed—that invaded the landscape, crowding out the native plants that had been there before them. Because these big canopy forests made the ground too dark for other trees to survive. And since there were frequent fires, the closed-canopy forests were able to spread like mad. So they're not timeless, Ted. They're merely the last in line."

Bradley snorted. "They're still 6 thousand years old..."

But Jennifer was relentless. "Not true," she said. "Scientists have shown that the forests continuously changed their composition. Each thousand-year period was different from the one before it. The forests changed constantly, Ted. And then, of course, there were the Indians."

"What about them?" ...

"The Indians were expert observers of the natural world, so they realized that old-growth forests sucked. Those forests may look impressive, but they're dead landscapes for game. So the Indians set fires, making sure the forests burned down periodically. They made sure there were only islands of old-growth forest in the midst of plains and meadows. The forests that the first Europeans saw were hardly primeval. They were cultivated, Ted. And it's not surprising that one hundred fifty years ago, there was less old-growth forest than there is today. The Indians were realists. Today, it's all romantic mythology." ...

After a while, Bradley excused himself and went to the front of the plane to call his agent. Evans said to Jennifer, "How did you know all that stuff?"

"For the reason Bradley himself mentioned. The 'dire threat of global warming.' We had a whole team researching dire threats. Because we wanted to find everything we could to make our case as impressive as possible."

"And?"

She shook her head. "The threat of global warming," she said, "is essentially nonexistent. Even if it were a real phenomenon, it would probably result in a net benefit."

APPENDIX C

CLIMATEGATE: THE CRU CONTROVERSY

Raymond Bradley
Currently a Professor in the Department of Geosciences and Director of the Climate System Research Center at the University of Massachusetts Amherst. Served as a Contributing Author in both the IPCC Third and Second Assessment Report.

Keith Briffa

Deputy Director of the Climatic Research Unit, University of East Anglia. Served as a Lead Author of the IPCC Fourth Assessment Report, a Contributing Author and Reviewer of the IPCC Third Assessment Report, and a Contributing Author of the IPCC Second Assessment Report.

Timothy Carter

Research Professor at the Finnish Environment Institute (SYKE), Helsinki, Finland. Served as an Expert Reviewer of the IPCC Fourth Assessment Report, Lead Author and Reviewer of the IPCC Third Assessment Report, and Convening Lead Author of the IPCC Second Assessment Report.

Edward Cook

Doherty Senior Scholar at the Tree-Ring Laboratory, Lamont-Doherty Earth Observatory, Palisades, New York. Served as a Contributing Author in the IPCC Fourth, Third, and Second Assessment Reports.

Malcolm Hughes

Regents' Professor in the Laboratory of Tree-Ring Research at the University of Arizona. Served as a Contributing Author and Reviewer of the IPCC Third Assessment Report.

Dr. Phil Jones

Professor at University of East Anglia's CRU. Served as a Coordinating Lead Author in the 2007 IPCC Fourth Assessment Report as well as an Expert Reviewer. Also was a Contributing Author in both the IPCC Third and IPCC Second Assessment Reports.

Thomas Karl

Designated Transitional Director of the NOAA Climate Service. Served as a Review Editor of the IPCC Fourth Assessment Report, Coordinating Lead Author and Lead Author of the IPCC Third Assessment Report, and both Lead and Contributing Author on the IPCC Second Assessment Report. Also has worked on multiple United States Global Change Research Programs (USGCRP) including his work as a Co-Chair and Synthesis Team Member of the USGCRP's 2000 U.S. National Assessment and Co-Chair and one of three Editors in Chief of the USGCRP's 2009 Global Climate Change Impacts in the United States Report. Also served as an Editor, Convening Lead Author, and Author of the USGCRP's 2008 Weather and Climate Extremes in a Changing Climate Report. Was Chief Editor and Federal Executive Team Member of the United States Climate Change Science Program's 2006 Temperature Trends in the Lower Atmosphere report.

Dr. Michael Mann

Professor and Director of Pennsylvania State University's Earth System Science Center. Served as an Expert Reviewer of the IPCC Fourth Assessment Report as well as a Lead Author, Contributing Author, and Reviewer of the IPCC Third Assessment Report.

Dr. Michael Oppenheimer

Albert G. Milbank Professor of Geosciences and International Affairs in the Woodrow Wilson School and the Department of Geosciences at Princeton University. Also is the Director of the Program in Science, Technology and Environmental Policy (STEP) at the Woodrow Wilson School and Faculty Associate of the Atmospheric and Ocean Sciences Program, Princeton Environmental Institute, and the Princeton Institute for International and Regional Studies. Served as a Lead Author, Contributing Author, and Expert Reviewer of the IPCC Fourth Report; Lead Author, Contributing Author, and Reviewer of the IPCC Third Assessment Report; and Contributing Author and Technical Summary Author of the IPCC Second Assessment Report.

Dr. Jonathan Overpeck

Co-Director of the Institute of the Environment as well as a Professor in the Department of Geosciences and the Department of Atmospheric Sciences at the University of Arizona. Served as a Coordinating Lead Author, Contributing Author, and Expert Reviewer of the IPCC Fourth Assessment Report; and Contributing Author of the IPCC Third and Second Assessment Reports.

Dr. Benjamin Santer
Research Scientist for the Program for Climate Model Diagnosis and Intercomparison at the Lawrence Livermore National Laboratory. Served as a Contributing Author in both the IPCC Fourth and Third Assessment Reports as well as Convening Lead Author, Technical Summary and Contributing Author of the IPCC Second Assessment Report. Also served as a Convening Lead Author, Lead Author, and Contributing Author in the U.S. CCSP's 2006 Temperature Trends in the Lower Atmosphere Report and Author of the USGCRP's 2009 Global Climate Change Impacts in the United States Report.

Gavin Schmidt
working at NASA's Goddard Institute for Space Studies. Served as a Contributing Author and Expert Reviewer for the IPCC Fourth Assessment Report.

Dr. Stephen Schneider
Melvin and Joan Lane Professor for Interdisciplinary Environmental Studies, Professor of Biological Sciences, Professor (by courtesy) of Civil and Environmental Engineering, and a Senior Fellow in the Woods Institute for the Environment at Stanford University. Served as a Reviewer of the IPCC Third Assessment Report and a Lead Author of the IPCC Second Assessment Report.

Dr. Susan Solomon
Senior Scientist at the Chemical Sciences Division (CSD) Earth System Research Laboratory (ESRL), NOAA. Served as a Co-Chair of the IPCC Working Group I, Contributing Author of the IPCC Fourth Assessment Report, and a Lead Author of the IPCC Third Assessment Report.

Peter Stott
Climate Monitoring Expert and Head of Climate Monitoring and Attribution at the Met Office Hadley Centre. Served as a Lead Author, Contributing Author, and Expert Reviewer of the IPCC Fourth Assessment Report and as a Contributing Author and Reviewer of the IPCC Third Assessment Report.

Dr. Kevin Trenberth
Senior Scientist and Head of the Climate Analysis Section at the National Center for Atmospheric Research. Served as a Coordinating Lead Author, Contributing Author, and Expert Reviewer of the IPCC Fourth Assessment Report; Lead Author, Contributing Author, and Reviewer of the IPCC Third Assessment Report; and Convening Lead Author, Technical Summary Author, and Contributing Author of the IPCC Second Assessment Report.

Dr. Thomas Wigley
Senior Scientist in the Climate and Global Dynamics Division, University Corporation for Atmospheric Research. Served as a Contributing Author of the IPCC Fourth and Third Assessment Reports as well as a Lead Author and Contributing Author of the IPCC Second Assessment Report. Also was a Convening Lead Author and Contributing Author of U.S. CCSP's 2006 Temperature Trends in the Lower Atmosphere report.

A SAMPLING OF EMAILS AND DOCUMENTS

The following is a preliminary sampling of CRU emails and documents which I believe seriously compromise the IPCC-backed "consensus" and its central conclusion that anthropogenic emissions are inexorably leading to environmental catastrophes, and which represent unethical and possibly illegal conduct by top IPCC scientists, among others. In the interest of brevity, many of the emails are not reproduced in their entirety. Therefore, the reader is encouraged to seek outside sources for broader review and context of the exposed emails and documents. The emails are reproduced in chronological order from oldest to newest on a variety of topics related to the participant's global warming research. These emails have been widely circulated:

> From: Michael E. Mann [University of Virginia]
> To: Tim Osborn [CRU]
> July 31, 2003
> Subject: Re: reconstruction errors
> Tim,
> Attached are the calibration residual series for experiments based on

available networks back to:

AD 1000

AD 1400

AD 1600

I can't find the one for the network back to 1820! But basically, you'll see that the residuals are pretty red for the first 2 cases, and then not significantly red for the 3rd case--its even a bit better for the AD 1700 and 1820 cases, but I can't seem to dig them up. . . . p.s. I know I probably don't need to mention this, but just to insure absolutely clarify on this, I'm providing these for your own personal use, since you're a trusted colleague. So please don't pass this along to others without checking w/ me first. This is the sort of "dirty laundry" one doesn't want to fall into the hands of those who might potentially try to distort things...

From: Phil Jones [CRU]
To: Michael E. Mann [University of Virginia]
January 16, 2004
Subject: CLIMATIC CHANGE needs your advice—YOUR EYES ONLY !!!!!
Mike,
This is for YOURS EYES ONLY. Delete after reading—please ! I'm trying to redress the balance. One reply from Pfister said you should make all available !! Pot calling the kettle black—Christian doesn't make his methods available. I replied to the wrong Christian message so you don't get to see what he said. Probably best. Told Steve separately and to get more advice from a few others as well as Kluwer and legal. PLEASE DELETE—just for you, not even Ray and Malcolm

From: Phil Jones
To: Tas van Ommen [University of Tasmania, Australia]
Cc: Michael E. Mann [University of Virginia]
February 9, 2004
Subject: Re: FW: Law Dome O18
Dear Tas,
Thanks for the email. Steve McIntyre hasn't contacted me directly about Law Dome (yet), nor about any of the series used in the 1998 Holocene paper or the 2003 GRL one with Mike. I suspect (hope) that he won't. I had some emails with him a few years ago when he wanted to get all the station temperature data we use here in CRU. I hid behind the fact that some of the data had been received from individuals and not directly from Met Services through the Global Telecommunications Service (GTS) or through GCOS. I've cc'd Mike on this, just for

info. Emails have also been sent to some other paleo people asking for datasets used in 1998 or 2003. Keith Briffa here got one, for example. Here, they have also been in contact with some of Keith's Russian contacts. All seem to relate to trying to get series we've used.

From: Michael E. Mann [University of Virginia]
To: Phil Jones [CRU]; Gabi Hergerl [Duke University]
August ??, 2004
[Subject: Mann and Jones (2003)]
Dear Phil and Gabi,
I've attached a cleaned-up and commented version of the matlab code that I wrote for doing the Mann and Jones (2003) composites. I did this knowing that Phil and I are likely to have to respond to more crap criticisms from the idiots in the near future, so best to clean up the code and provide to some of my close colleagues in case they want to test it, etc. Please feel free to use this code for your own internal purposes, but don't pass it along where it may get into the hands of the wrong people. . . .

From: Tom Wigley [University Corporation of Atmospheric Research]
To: Phil Jones [CRU]
January 21, 2005
Phil,
Thanks for the quick reply. The leaflet appeared so general, but it was prepared by UEA so they may have simplified things. From their wording, computer code would be covered by the FOIA. My concern was if Sarah is/was still employed by UEA. I guess she could claim that she had only written one tenth of the code and release every tenth line. Sorry I won't see you, but I will not come up to Norwich until Monday.

From: Phil Jones [CRU]
To: Tom Wigley [University Corporation of Atmospheric Research]
Cc: Ben Santer [Lawrence Livermore National Laboratory]
January 21st, 2005
Subject: Re: FOIA
Tom,
. . . As for FOIA Sarah isn't technically employed by UEA [University of East Anglia] and she will likely be paid by Manchester Metropolitan University. I wouldn't worry about the code. If FOIA does ever get used by anyone, there is also IPR [intellectual property rights] to consider as well. Data is covered by all the agreements we sign with people, so I will be hiding behind them. I'll be passing any requests onto the person

at UEA who has been given a post to deal with them.

From: Phil Jones [CRU]
To: Michael E. Mann [University of Virginia]
February 2, 2005
[Subject: For your eyes only]
Mike,

I presume congratulations are in order—so congrats etc ! Just sent loads of station data to Scott. Make sure he documents everything better this time ! And don't leave stuff lying around on ftp [file transfer protocol] sites—you never know who is trawling them. The two MMs have been after the CRU station data for years. If they ever hear there is a Freedom of Information Act now in the UK, I think I'll delete the file rather than send to anyone. Does your similar act in the US force you to respond to enquiries within 20 days?—our does ! The UK works on precedents, so the first request will test it. We also have a data protection act, which I will hide behind. Tom Wigley has sent me a worried email when he heard about it—thought people could ask him for his model code. He has retired officially from UEA so he can hide behind that. IPR [intellectual property rights] should be relevant here, but I can see me getting into an argument with someone at UEA who'll say we must adhere to it !

From: Michael E. Mann [University of Virginia]
To: Phil Jones [CRU]
February 2, 2005
Thanks Phil,
Yes, we've learned out lesson about FTP. We're going to be very careful in the future what gets put there. Scott really screwed up big time when he established that directory so that Tim could access the data. Yeah, there is a freedom of information act in the U.S., and the contrarians are going to try to use it for all its worth. But there are also intellectual property rights issues, so it isn't clear how these sorts of things will play out ultimately in the U.S. I saw the paleo draft (actually I saw an early version, and sent Keith some minor comments). It looks very good at present—will be interesting to see how they deal w/ the contrarian criticisms—there will be many. I'm hoping they'll stand firm (I believe they will—I think the chapter has the right sort of personalities for that)...

From: Phil Jones [CRU]
To: Michael E. Mann [University of Virginia]
Cc: Raymond Bradley [University of Massachusetts, Amherst]; Mal-

colm Hughes [University of Arizona]
February 21, 2005
Subject: Fwd: CCNet: PRESSURE GROWING ON CONTROVER-
SIAL RESEARCHER TO DISCLOSE SECRET DATA
Mike, Ray and Malcolm,
The skeptics seem to be building up a head of steam here ! Maybe we
can use this to our advantage to get the series updated ! Odd idea to
update the proxies with satellite estimates of the lower troposphere
rather than surface data !. Odder still that they don't realise that
Moberg et al used the Jones and Moberg updated series ! Francis Zwiers
is till onside. He said that PC1s produce hockey sticks. He stressed that
the late 20th century is the warmest of the millennium, but Regaldo
didn't bother with that. Also ignored Francis' comment about all the
other series looking similar to MBH [Mann Bradley Hughes]. The
IPCC comes in for a lot of stick. Leave it to you to delete as appropriate!
Cheers
Phil
PS I'm getting hassled by a couple of people to release the CRU station
temperature data.
Don't any of you three tell anybody that the UK has a Freedom of
Information Act !

From: Phil Jones [CRU]
To: Eugene R. Wahl [Alfred University]; Caspar Ammann [University
Corporation of Atmospheric Research]
September 12, 2007
Subject: Wahl/Ammann
Gene/Caspar,
Good to see these two out. Wahl/Ammann doesn't appear to be in CC's
online first, but comes up if you search. You likely know that McIntyre
will check this one to make sure it hasn't changed since the IPCC
close-off date July 2006! Hard copies of the WG1 report from CUP
have arrived here today. Ammann/Wahl—try and change the Received
date! Don't give those skeptics something to amuse themselves with.

From: Phil Jones [CRU]
To: Michael E. Mann [Penn State University]
May 29, 2008
Subject: IPCC & FOI
Mike,
Can you delete any emails you may have had with Keith re AR4 [IPCC
Fourth Assessment Report]? Keith will do likewise. He's not in at the

moment—minor family crisis. Can you also email Gene and get him to do the same? I don't have his new email address. We will be getting Caspar to do likewise. I see that CA [Climate Audit website] claim they discovered the 1945 problem in the Nature paper!!
Cheers
Phil

From: Michael E. Mann [Penn State University]
To: Phil Jones [CRU]
May 29, 2008
Subject: Re: IPCC & FOI
Hi Phil,
laughable that CA [Climate Audit] would claim to have discovered the problem. They would have run off to the Wall Street Journal for an exclusive were that to have been true. I'll contact Gene about this [deleting emails] ASAP. His new email is: . . .
talk to you later,
Mike

From: Phil Jones [CRU]
To: Gavin Schmidt [NASA Goddard Institute for Space Studies]
Cc: Michael E. Mann [Penn State University]
August 20, 2008
Gavin,
. . . Thinking about the final bit for the Appendix. Keith should be in later, so I'll check with him—and look at that vineyard book. I did rephrase the bit about the 'evidence' as Lamb refers to it. I wanted to use his phrasing—he used this word several times in these various papers. What he means is his mind and its inherent bias(es). Your final sentence though about improvements in reviewing and traceability is a bit of a hostage to fortune. The skeptics will try to hang on to something, but I don't want to give them something clearly tangible. Keith/Tim still getting FOI requests as well as MOHC [Meteorological Office Hadley Center] and Reading. All our FOI officers have been in discussions and are now using the same exceptions not to respond—advice they got from the Information Commissioner. . . . The FOI line we're all using is this. IPCC is exempt from any countries FOI—the skeptics have been told this. Even though we (MOHC, CRU/UEA) possibly hold relevant info the IPCC is not part our remit (mission statement, aims etc) therefore we don't have an obligation to pass it on.

From: Phil Jones [CRU]

To: Unknown list
March 10, 2003
[Subject: Soon & Baliunas]
Dear all,
Tim Osborn has just come across this. Best to ignore probably, so don't let it spoil your day. I've not looked at it yet. It results from this journal having a number of editors. The responsible one for this is a well-known skeptic in NZ. He has let a few papers through by Michaels and Gray in the past. I've had words with Hans von Storch about this, but got nowhere. Another thing to discuss in Nice !
Cheers
Phil

From: Phil Jones
To: Raymond Bradley [University of Massachusetts, Amherst]; Malcolm Hughes [University of Arizona]; Scott Rutherford [University of Rhode Island]; Michael E. Mann [University of Virginia]; Tom Crowley [Duke University]
Cc: Keith Briffa [CRU]; Jonathan Overpeck [University of Arizona]; Edward Cook [Columbia University]; Keith Alverson [IGBP-PAGES]
March 11, 2003
Subject: Fwd: Soon & Baliunas
Dear All,
Apologies for sending this again. I was expecting a stack of emails this morning in response, but I inadvertently left Mike off (mistake in pasting) and picked up Tom's old address. Tom is busy though with another offspring ! I looked briefly at the paper last night and it is appalling—worst word I can think of today without the mood pepper appearing on the email ! I'll have time to read more at the weekend as I'm coming to the US for the DoE CCPP meeting at Charleston. Added Ed, Peck and Keith A. onto this list as well. I would like to have time to rise to the bait, but I have so much else on at the moment. As a few of us will be at the EGS/AGU meet in Nice, we should consider what to do there. The phrasing of the questions at the start of the paper determine the answer they get. They have no idea what multiproxy averaging does. By their logic, I could argue 1998 wasn't the warmest year globally, because it wasn't the warmest everywhere. With their LIA [Little Ice Age] being 1300-1900 and their MWP [Medieval Warm Period] 800-1300, there appears (at my quick first reading) no discussion of synchroneity of the cool/warm periods. Even with the instrumental record, the early and late 20th century warming periods are only significant locally at between 10-20% of grid boxes. Writing

this I am becoming more convinced we should do something—even if this is just to state once and for all what we mean by the LIA and MWP. I think the skeptics will use this paper to their own ends and it will set paleo[climatology] back a number of years if it goes unchallenged. I will be emailing the journal to tell them I'm having nothing more to do with it until they rid themselves of this troublesome editor. A CRU person is on the editorial board, but papers get dealt with by the editor assigned by Hans von Storch.

Cheers

Phil

From: Michael E. Mann [University of Virginia]
To: Phil Jones [CRU]; Raymond Bradley [University of Massachusetts, Amherst]; Malcolm Hughes [University of Arizona]; Scott Rutherford [University of Rhode Island]; Tom Crowley [Duke University]
Cc: Keith Briffa [CRU]; Jonathan Overpeck [University of Arizona]; Edward Cook [Columbia University]; Keith Alverson [IGBP-PAGES]; Mike MacCracken [Climate Institute]
March 11, 2003
Subject: Re: Fwd: Soon & Baliunas
Thanks Phil,
(Tom: Congrats again!)
The Soon & Baliunas paper couldn't have cleared a 'legitimate' peer review process anywhere. That leaves only one possibility—that the peer-review process at Climate Research has been hijacked by a few skeptics on the editorial board. And it isn't just De Frietas, unfortunately I think this group also includes a member of my own department... The skeptics appear to have staged a 'coup' at "Climate Research" (it was a mediocre journal to begin with, but now its a mediocre journal with a definite 'purpose'). Folks might want to check out the editors and review editors: [1]http://www.int-res.com/journals/cr/crEditors.html In fact, Mike McCracken first pointed out this article to me, and he and I have discussed this a bit. I've cc'd Mike in on this as well, and I've included Peck too. I told Mike that I believed our only choice was to ignore this paper. They've already achieved what they wanted—the claim of a peer-reviewed paper. There is nothing we can do about that now, but the last thing we want to do is bring attention to this paper, which will be ignored by the community on the whole... It is pretty clear that thee skeptics here have staged a bit of a coup, even in the presence of a number of reasonable folks on the editorial board (Whetton, Goodess, ...). My guess is that Von Storch is actually with them (frankly, he's an odd individual, and I'm not sure he isn't himself

somewhat of a skeptic himself), and without Von Storch on their side, they would have a very forceful personality promoting their new vision. There have been several papers by Pat Michaels, as well as the Soon & Baliunas paper, that couldn't get published in a reputable journal. This was the danger of always criticising the skeptics for not publishing in the "peer-reviewed literature". Obviously, they found a solution to that--take over a journal! So what do we do about this? I think we have to stop considering "Climate Research" as a legitimate peer-reviewed journal. Perhaps we should encourage our colleagues in the climate research community to no longer submit to, or cite papers in, this journal. We would also need to consider what we tell or request of our more reasonable colleagues who currently sit on the editorial board... What do others think?
Mike

From: Michael E. Mann [University of Virginia]
To: Malcolm Hughes [University of Arizona]
March 11, 2003
HI Malcolm,
Thanks for the feedback--I largely concur. I do, though, think there is a particular problem with "Climate Research". This is where my colleague Pat Michaels now publishes exclusively, and his two closest colleagues are on the editorial board and review editor board. So I promise you, we'll see more of this there, and I personally think there *is* a bigger problem with the "messenger" in this case...

From: Phil Jones [CRU]
To: Unknown List
March 12, 2003
Dear All,
I agree with all the points being made and the multi-authored article would be a good idea, but how do we go about not letting it get buried somewhere. Can we not address the misconceptions by finally coming up with definitive dates for the LIA and MWP and redefining what we think the terms really mean? With all of us and more on the paper, it should carry a lot of weight. In a way we will be setting the agenda for what should be being done over the next few years...

From: Tom Wigley [University Corporation of Atmospheric Research]
To: Phil Jones [CRU]; Keith Briffa [CRU]; James Hansen [NASA Goddard Institute for Space Studies]; Michael E. Mann [University of Virginia]; Ben Santer [Lawrence Livermore National Laboratory];

Thomas R Karl [NOAA]; Mark Eakin [NOAA]; et al.
April 23, 2003
Subject: My turn

. . . This second case gets to the crux of the matter. I suspect that deFreitas deliberately chose other referees who are members of the skeptics camp. I also suspect that he has done this on other occasions. How to deal with this is unclear, since there are a number of individuals with bona fide scientific credentials who could be used by an unscrupulous editor to ensure that 'anti-greenhouse' science can get through the peer review process (Legates, Balling, Lindzen, Baliunas, Soon, and so on). The peer review process is being abused, but proving this would be difficult. The best response is, I strongly believe, to rebut the bad science that does get through. Jim Salinger raises the more personal issue of deFreitas. He is clearly giving good science a bad name, but I do not think a barrage of ad hominem attacks or letters is the best way to counter this. If Jim wishes to write a letter with multiple authors, I may be willing to sign it, but I would not write such a letter myself. In this case, deFreitas is such a poor scientist that he may simply disappear. I saw some work from his PhD, and it was awful (Pat Michaels' PhD is at the same level).
Best wishes to all,
Tom.

From: Mark Eakin [NOAA]
To: Michael E. Mann [University of Virginia]; et al.
April 24th, 2003
[Subject: My turn]

. . . A letter to OSTP [White House Office of Science and Technology Policy] is probably in order here. Since the White House has shown interest in this paper, OSTP really does need to receive a measured, critical discussion of flaws in Soon and Baliunas' methods. I agree with Tom that a noted group from the detection and attribution effort such as Mann, Crowley, Briffa, Bradley, Jones and Hughes should spearhead such a letter. Many others of us could sign on in support. This would provide Dave Halpern with the ammunition he needs to provide the White House with the needed documentation that hopefully will dismiss this paper for the slipshod work that it is. Such a letter could be developed in parallel with a rebuttal article...

From: Timothy Carter [Finnish Environment Institute]
To: Tom Wigley [University Corporation of Atmospheric Research]
April ??, 2003

[Subject: Java climate model]

. . . P.S. On the CR [Climate Research] issue, I agree that a rebuttal seems to be the only method of addressing the problem (I communicated this to Mike yesterday morning), and I wonder if a review of the refereeing policy is in order. The only way I can think of would be for all papers to go through two Editors rather than one, the former to have overall responsibility, the latter to provide a second opinion on a paper and reviewers' comments prior to publication. A General Editor would be needed to adjudicate in the event of disagreement. Of course, this could then slow down the review process enormously. However, without an editorial board to vote someone off, how can suspect Editors be removed except by the Publisher (in this case, Inter-Research).

From: Tom Wigley [University Corporation of Atmospheric Research]
To: Timothy Carter [Finnish Environment Institute]
Cc: Mike Hulme [CRU]; Phil Jones [CRU]
April 24, 2003
Subject: Re: Java climate model

. . . PS Re CR, I do not know the best way to handle the specifics of the editoring. Hans von Storch is partly to blame—he encourages the publication of crap science 'in order to stimulate debate'. One approach is to go direct to the publishers and point out the fact that their journal is perceived as being a medium for disseminating misinformation under the guise of refereed work. I use the word 'perceived' here, since whether it is true or not is not what the publishers care about—it is how the journal is seen by the community that counts. I think we could get a large group of highly credentialed scientists to sign such a letter—50+ people. Note that I am copying this view only to Mike Hulme and Phil Jones. Mike's idea to get editorial board members to resign will probably not work—must get rid of von Storch too, otherwise holes will eventually fill up with people like Legates, Balling, Lindzen, Michaels, Singer, etc. I have heard that the publishers are not happy with von Storch, so the above approach might remove that hurdle too.

From: Edward Cook [Columbia University]
To: Keith Briffa [CRU]
June 4, 2003
[Subject: Review- confidential REALLY URGENT]
Hi Keith,

Okay, today. Promise! Now something to ask from you. Actually somewhat important too. I got a paper to review (submitted to the Journal of Agricultural, Biological, and Environmental Sciences), written by a

Korean guy and someone from Berkeley, that claims that the method of reconstruction that we use in dendroclimatology (reverse regression) is wrong, biased, lousy, horrible, etc. They use your Tornetrask recon as the main whipping boy. . . . I would like to play with it in an effort to refute their claims. If published as is, this paper could really do some damage. It is also an ugly paper to review because it is rather mathematical, with a lot of Box-Jenkins stuff in it. It won't be easy to dismiss out of hand as the math appears to be correct theoretically but it suffers from the classic problem of pointing out theoretical deficiencies . . . I am really sorry but I have to nag about that review—Confidentially I now need a hard and if required extensive case for rejecting—to support Dave Stahle's and really as soon as you can.

From: Andrew Comrie [University of Arizona]
To: Phil Jones [CRU]
May, 2004
[Subject: IJOC040512 review]
Dear Prof. Jones,
IJOC040512 "A Socioeconomic Fingerprint on the Spatial Distribution of Surface Air Temperature Trends"
Authors: RR McKitrick & PJ Michaels
Target review date: July 5, 2004
I know you are very busy, but do you have the time to review the above manuscript [from skeptics McKitrick and Michaels] for the International Journal of Climatology? If yes, can you complete the review within about five to six weeks, say by the target review date listed above? I will send the manuscript electronically. If no, can you recommend someone who you think might be a good choice to review this paper? . . .
[Note: In the peer review process, reviewer's names are kept anonymous.]

From: Phil Jones [CRU]
To: Andrew Comrie [University of Arizona]
May 24, 2004
Subject: RE: IJOC040512 review
Andrew,
I can do this. I am in France this week but back in the UK all June. So send and it will be waiting my return.
Phil

From: Phil Jones [CRU]
To: Michael E. Mann [University of Virginia]
August 13, 2004
Subject: Fwd: RE: IJOC040512 review
Mike,
The paper ! Now to find my review. I did suggest to Andrew to find
3 reviewers.
Phil

From: Michael E. Mann [University of Virginia]
To: Phil Jones [CRU]
August 13, 2004
[Subject: IJOC040512 review]
Thanks a bunch Phil,
Along lines as my other email, would it be (?) for me to forward this to
the chair of our commitee confidentially, and for his internal purposes
only, to help bolster the case against MM [skeptics McKitrick and
Michaels]?? let me know...
thanks,
Mike

From: Phil Jones [CRU]
To: Michael E. Mann [University of Virginia]
August 13, 2004
Subject: Re: Fwd: RE: IJOC040512 review
Mike,
I'd rather you didn't. I think it should be sufficient to forward the
para from Andrew Conrie's email that says the paper has been rejected
by all 3 reviewers. You can say that the paper was an extended and
updated version of that which appeared in CR. Obviously, under no
circumstances should any of this get back to Pielke.
Cheers
Phil

From: Phil Jones [CRU]
To: Michael E. Mann [University of Virginia]
July 8, 2004
Subject: HIGHLY CONFIDENTIAL
Mike,
Only have it in the pdf form. FYI ONLY—don't pass on. Relevant
paras are the last 2 in section 4 on p13. As I said it is worded carefully
due to Adrian knowing Eugenia for years. He knows the're wrong, but

he succumbed to her almost pleading with him to tone it down as it might affect her proposals in the future ! I didn't say any of this, so be careful how you use it—if at all. Keep quiet also that you have the pdf. The attachment is a very good paper—I've been pushing Adrian over the last weeks to get it submitted to JGR [Journal of Geophysical Research] or J. Climate [Journal of Climate]. The main results are great for CRU and also for ERA-40. The basic message is clear—you have to put enough surface and sonde obs into a model to produce Reanalyses. The jumps when the data input change stand out so clearly. NCEP does many odd things also around sea ice and over snow and ice. . . . The other paper by MM is just garbage—as you knew. De Freitas again. Pielke is also losing all credibility as well by replying to the mad Finn as well—frequently as I see it. I can't see either of these papers being in the next IPCC report. Kevin and I will keep them out somehow—even if we have to redefine what the peer-review literature is!
Cheers
Phil

Mike,

For your interest, there is an ECMWF ERA-40 Report coming out soon, which shows that Kalnay and Cai are wrong. It isn't that strongly worded as the first author is a personal friend of Eugenia. The result is rather hidden in the middle of the report. It isn't peer review, but a slimmed down version will go to a journal. KC are wrong because the difference between NCEP and real surface temps (CRU) over eastern N. America doesn't happen with ERA-40. ERA-40 assimilates surface temps (which NCEP didn't) and doing this makes the agreement with CRU better. Also ERA-40's trends in the lower atmosphere are all physically consistent where NCEP's are not—over eastern US. I can send if you want, but it won't be out as a report for a couple of months.
Cheers
Phil

From: Stephen Mackwell [Universities Space Research Association]
To: Michael E. Mann [University of Virginia]
Cc: Chris Reason [University of Cape Town]; James Saiers [Yale University]
January 20, 2005
Subject: Your concerns with 2004GL021750 McIntyre
Dear Prof. Mann
In your recent email to Chris Reason, you laid out your concerns that I presume were the reason for your phone call to me last week. I have reviewed the manuscript by McIntyre, as well as the reviews. The

editor in this case was Prof. James Saiers. He did note initially that the manuscript did challenge published work, and so felt the need for an extensive and thorough review. For that reason, he requested reviews from 3 knowledgable scientists. All three reviews recommended publication. While I do agree that this manuscript does challenge (somewhat aggresively) some of your past work, I do not feel that it takes a particularly harsh tone. On the other hand, I can understand your reaction. As this manuscript was not written as a Comment, but rather as a full-up scientific manuscript, you would not in general be asked to look it over. And I am satisfied by the credentials of the reviewers. Thus, I do not feel that we have sufficient reason to interfere in the timely publication of this work. However, you are perfectly in your rights to write a Comment, in which you challenge the authors' arguments and assertions. Should you elect to do this, your Comment would be provided to them and they would be offered the chance to write a Reply. Both Comment and Reply would then be reviewed and published together (if they survived the review process). Comments are limited to the equivalent of 2 journal pages.

Regards

Steve Mackwell

Editor in Chief, GRL [Geophysical Research Letters]

From: Michael E. Mann [University of Virginia]
The following individuals may have been recipients: Tom Wigley [University Corporation of Atmospheric Research]; Raymond Bradley [University of Massachusetts, Amherst]; Tom Osborn [CRU]; Phil Jones [CRU]; Keith Briffa [CRU]; Gavin Schmidt [NASA Goddard Institute for Space Studies]; Malcolm Hughes [University of Arizona]; [Subject: Your concerns with 2004GL021750 McIntyre]
January 20, 2005
Dear All,

Just a heads up. Apparently, the contrarians now have an "in" with GRL [Geophysical Research Letters]. This guy Saiers has a prior connection w/ the University of Virginia Dept. of Environmental Sciences that causes me some unease. I think we now know how the various Douglass et al papers w/ Michaels and Singer, the Soon et al paper, and now this one have gotten published in GRL,

Mike

From: Tom Wigley [University Corporation of Atmospheric Research]
To: Michael E. Mann [University of Virginia]
The following individuals may also have been recipients: Raymond

Bradley [University of Massachusetts, Amherst]; Tom Osborn [CRU]; Phil Jones [CRU]; Keith Briffa [CRU]; Gavin Schmidt [NASA Goddard Institute for Space Studies]; Malcolm Hughes [University of Arizona];

January 20, 2005

[Subject: Your concerns with 2004GL021750 McIntyre]

Mike,

This is truly awful. GRL [Geophysical Research Letters] has gone downhill rapidly in recent years. I think the decline began before Saiers. I have had some unhelpful dealings with him recently with regard to a paper Sarah and I have on glaciers—it was well received by the referees, and so is in the publication pipeline. However, I got the impression that Saiers was trying to keep it from being published. Proving bad behavior here is very difficult. If you think that Saiers is in the greenhouse skeptics camp, then, if we can find documentary evidence of this, we could go through official AGU [American Geophysical Union] channels to get him ousted.

From: Michael E. Mann [University of Virginia]

To: Tom Wigley [University Corporation of Atmospheric Research]

The following individuals may also have been recipients: Raymond Bradley [University of Massachusetts, Amherst]; Tom Osborn [CRU]; Phil Jones [CRU]; Keith Briffa [CRU]; Gavin Schmidt [NASA Goddard Institute for Space Studies]; Malcolm Hughes [University of Arizona];

January 20, 2005

[Subject: Your concerns with 2004GL021750 McIntyre]

Thanks Tom,

Yeah, basically this is just a heads up to people that something might be up here. What a shame that would be. It's one thing to lose "Climate Research". We can't afford to lose GRL [Geophysical Research Letters]. I think it would be useful if people begin to record their experiences w/ both Saiers and potentially Mackwell (I don't know him--he would seem to be complicit w/ what is going on here). If there is a clear body of evidence that something is amiss, it could be taken through the proper channels. I don't that the entire AGU [American Geophysical Union] hierarchy has yet been compromised! The GRL article simply parrots the rejected Nature comment--little substantial difference that I can see at all. Will keep you all posted of any relevant developments, Mike

From: Michael E. Mann [University of Virginia]

To: Malcolm Hughes [University of Arizona]
The following individuals may also have been recipients: Tom Wigley [University Corporation of Atmospheric Research]; Raymond Bradley [University of Massachusetts, Amherst]; Tom Osborn [CRU]; Phil Jones [CRU]; Keith Briffa [CRU]; Gavin Schmidt [NASA Goddard Institute for Space Studies]; Malcolm Hughes [University of Arizona]; January 20 or 21, 2005
[Subject: Your concerns with 2004GL021750 McIntyre]
Hi Malcolm,
This assumes that the editor/s in question would act in good faith. I'm not convinced of this. I don't believe a response in GRL is warranted in any case. The MM claims in question are debunked in other papers that are in press and in review elsewhere. I'm not sure that GRL can be seen as an honest broker in these debates anymore, and it is probably best to do an end run around GRL now where possible. They have published far too many deeply flawed contrarian papers in the past year or so. There is no possible excuse for them publishing all 3 Douglass papers and the Soon et al paper. These were all pure crap. There appears to be a more fundamental problem w/ GRL now, unfortunately...
Mike

From: Ben Santer [Lawrence Livermore National Laboratory]
To: Phil Jones [CRU]
March 19, 2009
[Subject: See the link below]
. . . If the RMS [Royal Meteorological Society] is going to require authors to make ALL data available—raw data PLUS results from all intermediate calculations—I will not submit any further papers to RMS journals.
Cheers,
Ben

From: Phil Jones [CRU]
To: Ben Santer [Lawrence Livermore National Laboratory]
March 19, 2009
Subject: Re: See the link below
. . . I'm having a dispute with the new editor of Weather. I've complained about him to the RMS Chief Exec. If I don't get him to back down, I won't be sending any more papers to any RMS journals and I'll be resigning from the RMS.

From: Kevin Trenberth [University Corporation of Atmospheric

Research]
To: Michael E. Mann [Penn State University]
Cc: Grant Foster; Phil Jones [CRU]; Gavin Schmidt [NASA Goddard Institute for Space Studies]; et al.
July 29, 2009
Subject: Re: ENSO blamed over warming—paper in JGR
Hi all
Wow this is a nice analysis by Grant et al. What we should do is turn this into a learning experience for everyone: there is often misuse of filtering. Obviously the editor and reviewers need to to also be taken to task here. I agree with Mike Mann that a couple of other key points deserve to be made wrt this paper. . . .

From: Keith Briffa [CRU]
To: Chris Folland [UK Met Office]; Phil Jones [CRU]; Michael E. Mann [University of Virginia]
Cc: Tom Karl [National Climatic Data Center—NOAA]
September 22, 1999
Subject: RE: IPCC revisions
. . . I know there is pressure to present a nice tidy story as regards 'apparent unprecedented warming in a thousand years or more in the proxy data' but in reality the situation is not quite so simple. We don't have a lot of proxies that come right up to date and those that do (at least a significant number of tree proxies) some unexpected changes in response that do not match the recent warming. . . .

From: Phil Jones [CRU]
To: Ray Bradley [University of Massachusetts, Amherst]; Michael E. Mann [University of Virginia]; Malcolm Hughes [University of Arizona]
Cc: Keith Briffa [CRU]; Tom Osborn [CRU]
November 16, 1999
Subject: Diagram for WMO [World Meteorological Organization] Statement
Dear Ray, Mike and Malcolm,
Once Tim's got a diagram here we'll send that either later today or first thing tomorrow. I've just completed Mike's Nature trick of adding in the real temps to each series for the last 20 years (ie from 1981 onwards) amd from 1961 for Keith's to hide the decline. Mike's series got the annual land and marine values while the other two got April-Sept for NH [Northern Hemisphere] land N of 20N. The latter two are real for 1999, while the estimate for 1999 for NH combined is +0.44C wrt

61-90. The Global estimate for 1999 with data through Oct is +0.35C cf. 0.57 for 1998. Thanks for the comments, Ray.
Cheers
Phil

From: Giorgi Filippo [International Centre for Theoretical Physics]
To: Chapter 10 LAs
September 11, 2000
Subject: On "what to do?"
Given this, I would like to add my own opinion developed through the weekend. First let me say that in general, as my own opinion, I feel rather unconfortable about using not only unpublished but also un reviewed material as the backbone of our conclusions (or any conclusions). I realize that chapter 9 is including SRES stuff, and thus we can and need to do that too, but the fact is that in doing so the rules of IPCC have been softened to the point that in this way the IPCC is not any more an assessment of published science (which is its proclaimed goal) but production of results. The softened condition that the models themself have to be published does not even apply because the Japanese model for example is very different from the published one which gave results not even close to the actual outlier version (in the old dataset the CCC model was the outlier). Essentially, I feel that at this point there are very little rules and almost anything goes. I think this will set a dangerous precedent which might mine the IPCC credibility, and I am a bit uncomfortable that now nearly everybody seems to think that it is just ok to do this. Anyways, this is only my opinion for what it is worth.

From: Michael E. Mann [University of Virginia]
To: Phil Jones [CRU]; et al.
June 4, 2003
Subject: Re: Prospective Eos piece?
. . . Phil and I have recently submitted a paper using about a dozen NH records that fit this category, and many of which are available nearly 2K back--I think that trying to adopt a timeframe of 2K, rather than the usual 1K, addresses a good earlier point that Peck made w/ regard to the memo, that it would be nice to try to "contain" the putative "MWP", even if we don't yet have a hemispheric mean reconstruction available that far back [Phil and I have one in review--not sure it is kosher to show that yet though--I've put in an inquiry to Judy Jacobs at AGU about this]. . . .

From: David Rind [NASA Goddard Institute for Space Studies]

To: Jonathan Overpeck [University of Arizona]
January 4, 2005
[Subject: IPCC last 2000 years data]
. . . In addition, some of the comments are probably wrong—the warm-season bias (p.12) should if anything produce less variability, since warm seasons (at least in GCMs) feature smaller climate changes than cold seasons. The discussion of uncertainties in tree ring reconstructions should be direct, not referred to other references—it's important for this document. How the long-term growth is factored in/out should be mentioned as a prime problem. The lack of tropical data—a few corals prior to 1700—has got to be discussed. The primary criticism of McIntyre and McKitrick, which has gotten a lot of play on the Internet, is that Mann et al. transformed each tree ring prior to calculating PCs by subtracting the 1902-1980 mean, rather than using the length of the full time series (e.g., 1400-1980), as is generally done. M&M claim that when they used that procedure with a red noise spectrum, it always resulted in a 'hockey stick'. Is this true? If so, it constitutes a devastating criticism of the approach; if not, it should be refuted. While IPCC cannot be expected to respond to every criticism a priori, this one has gotten such publicity it would be foolhardy to avoid it. . . .

From: Jonathan Overpeck [University of Arizona]
To: Keith Briffa [CRU]; Eystein Jansen [Bjerknes Centre for Climate Research]; Tom Crowley [Duke University]
July ??, 2005
ANOTHER THING THAT IS A REAL ISSUE IS SHOWING SOME OF THE TREE-RING DATA FOR THE PERIOD AFTER 1950. BASED ON THE LITERATURE, WE KNOW THESE ARE BIASED—RIGHT? SO SHOULD WE SAY THAT'S THE REASON THEY ARE NOT SHOWN? OF COURSE, IF WE ONLY PLOT THE FIG FROM CA 800 TO 1400 AD, IT WOULD DO WHAT WE WANT, FOCUS ON THE MWP ONLY—THE TOPIC OF THE BOX—AND SHOW THAT THERE WERE NOT ANY PERIODS WHEN ALL THE RECORDS ALL SHOWED WARMTH—I.E., OF THE KIND WE'RE EXPERIENCING NOW. TWO CENTS WORTH

From: Michael E. Mann [Penn State University]
To: Tim Osborn [CRU]; Keith Briffa [CRU]
Cc: Gavin Schmidt [NASA Goddard Institute for Space Studies]
February 9, 2006
guys, I see that Science has already gone online w/ the new issue, so we

put up the RC [Real Climate website] post. By now, you've probably read that nasty McIntyre thing. Apparently, he violated the embargo on his website (I don't go there personally, but so I'm informed). Anyway, I wanted you guys to know that you're free to use RC in any way you think would be helpful. Gavin and I are going to be careful about what comments we screen through, and we'll be very careful to answer any questions that come up to any extent we can. On the other hand, you might want to visit the thread and post replies yourself. We can hold comments up in the queue and contact you about whether or not you think they should be screened through or not, and if so, any comments you'd like us to include. You're also welcome to do a followup guest post, etc. think of RC as a resource that is at your disposal to combat any disinformation put forward by the McIntyres of the world. Just let us know. We'll use our best discretion to make sure the skeptics dont'get to use the RC comments as a megaphone...

From: Keith Briffa [CRU]
To: Martin Juckes [???]; et al.
November 16, 2006
Subject: Re: Mitrie: Bristlecones
. . . I still believe that it would be wise to involve Malcolm Hughes in this discussion—though I recognise the point of view that says we might like to appear (and be) independent of the original Mann, Bradley and Hughes team to avoid the appearance of collusion. In my opinion (as someone how has worked with the Bristlecone data hardly at all!) there are undoubtedly problems in their use that go beyond the strip bark problem (that I will come back to later). . . . Another serious issue to be considered relates to the fact that the PC1 time series in the Mann et al. analysis was adjusted to reduce the positive slope in the last 150 years (on the assumption—following an earlier paper by Lamarche et al.—that this incressing growth was evidence of carbon dioxide fertilization) , by differencing the data from another record produced by other workers in northern Alaska and Canada (which incidentally was standardised in a totally different way). This last adjustment obviously will have a large influence on the quantification of the link between these Western US trees and N.Hemisphere temperatures. At this point, it is fair to say that this adjustment was arbitrary and the link between Bristlecone pine growth and CO_2 is , at the very least, arguable.

From: Tom Wigley [University Corporation of Atmospheric Research]
To: Phil Jones [CRU]
Cc: Ben Santer [Lawrence Livermore National Laboratory]

September 27, 2009
Subject: 1940s
Phil,

Here are some speculations on correcting SSTs [Sea Surface Temperatures] to partly explain the 1940s warming blip. If you look at the attached plot you will see that the land also shows the 1940s blip (as I'm sure you know). So, if we could reduce the ocean blip by, say, 0.15 degC, then this would be significant for the global mean—but we'd still have to explain the land blip. I've chosen 0.15 here deliberately. This still leaves an ocean blip, and i think one needs to have some form of ocean blip to explain the land blip (via either some common forcing, or ocean forcing land, or vice versa, or all of these). When you look at other blips, the land blips are 1.5 to 2 times (roughly) the ocean blips—higher sensitivity plus thermal inertia effects. My 0.15 adjustment leaves things consistent with this, so you can see where I am coming from. . . .

From: Tom Wigley [University Corporation of Atmospheric Research]
To: Phil Jones [CRU]
October 5, 2009
[Subject: A Scientific Scandal Unfolds]
Phil,

It is distressing to read that American Stinker item [Oct. 5th article from the American Thinker which highlights Stephen McIntyre's discovery that Keith Briffa apparently cherry picked data regarding tree-rings from Yamal]. But Keith does seem to have got himself into a mess. As I pointed out in emails, Yamal is insignificant. And you say that (contrary to what M&M say) Yamal is *not* used in MBH, etc. So these facts alone are enough to shoot down M&M is a few sentences (which surely is the only way to go—complex and wordy responses will be counter productive). But, more generally, (even if it *is* irrelevant) how does Keith explain the McIntyre plot that compares Yamal-12 with Yamal-all? And how does he explain the apparent "selection" of the less well-replicated chronology rather that the later (better replicated) chronology? Of course, I don't know how often Yamal-12 has really been used in recent, post-1995, work. I suspect from what you say it is much less often that M&M say—but where did they get their information? I presume they went thru papers to see if Yamal was cited, a pretty foolproof method if you ask me. Perhaps these things can be explained clearly and concisely—but I am not sure Keith is able to do this as he is too close to the issue and probably quite pissed of. And the issue of with-holding data is still a hot potato, one that affects both

you and Keith (and Mann). Yes, there are reasons—but many *good* scientists appear to be unsympathetic to these. The trouble here is that with-holding data looks like hiding something, and hiding means (in some eyes) that it is bogus science that is being hidden. I think Keith needs to be very, very careful in how he handles this. I'd be willing to check over anything he puts together.
Tom.

From: Phil Jones [CRU]
To: John Mitchell [Director of Climate Science—UK Met Office]
October 28, 2009
Subject: Yamal response from Keith
John,
. . . This went up last night about 5pm. There is a lot to read at various levels. If you get time just the top level is necessary. There is also a bit from Tim Osborn showing that Yamal was used in 3 of the 12 millennial reconstructions used in Ch 6 [of IPCC Fourth Assessment Report]. Also McIntyre had the Yamal data in Feb 2004—although he seems to have forgotten this. Keith succeeding in being very restrained in his response. McIntyre knew what he was doing when he replaced some of the trees with those from another site.
Cheers
Phil

From: Phil Jones [CRU]
To: Keith Briffa [CRU]
October 28, 2009
Subject: FW: Yamal and paleoclimatology
Keith,
There is a lot more there on CA [Climate Audit website] now. I would be very wary about responding to this person now having seen what McIntyre has put up. You and Tim talked about Yamal. Why have the bristlecones come in now. . . . This is what happens—they just keep moving the goalposts. Maybe get Tim to redo OB2006 without a few more series.
Cheers
Phil . . .
Dear Professor Briffa, I am pleased to hear that you appear to have recovered from your recent illness sufficiently to post a response to the controversy surrounding the use of the Yamal chronology; and the chronology itself; Unfortunately I find your explanations lacking in scientific rigour and I am more inclined to believe the analysis of

McIntyre[.] Can I have a straightforward answer to the following questions 1) Are the reconstructions sensitive to the removal of either the Yamal data and Strip pine bristlecones, either when present singly or in combination? 2) Why these series, when incorporated with white noise as a background, can still produce a Hockey-Stick shaped graph if they have, as you suggest, a low individual weighting? And once you have done this, please do me the courtesy of answering my initial email. Dr. D.R. Keiller

From: Keith Briffa [CRU]
To: Chris Folland [UK Met Office]; Phil Jones [CRU]; Michael E. Mann [University of Virginia]
Cc: Tom Karl [National Climatic Data Center—NOAA]
September 22, 1999
Subject: RE: IPCC revisions
. . . I know there is pressure to present a nice tidy story as regards 'apparent unprecedented warming in a thousand years or more in the proxy data' but in reality the situation is not quite so simple. We don't have a lot of proxies that come right up to date and those that do (at least a significant number of tree proxies) some unexpected changes in response that do not match the recent warming. . . .

From: Edward Cook [Columbia University]
To: Keith Briffa [CRU]
April 29, 2003
[Subject: Review- confidential]
Hi Keith,
I will start out by sending you the chronologies that I sent Bradley, i.e. all but Mongolia. If you can talk Gordon out of the latter, you'll be the first from outside this lab. The chronologies are in tabbed column format and Tucson index format. The latter have sample size included. It doesn't take a rocket scientist (or even Bradley after I warned him about small sample size problems) to realize that some of the chronologies are down to only 1 series in their earliest parts. Perhaps I should have truncated them before using them, but I just took what Jan gave me and worked with the chronologies as best I could. My suspicion is that most of the pre-1200 divergence is due to low replication and a reduced number of available chronologies. I should also say that the column data have had their means normalized to approximately 1.0, which is not the case for the chronologies straight out of ARSTAN. That is because the site-level RCS-detrended data were simply aver-

aged to produce these chronologies, without concern for their long-term means. Hence the "RAW" tag at the end of each line of indices. Bradley still regards the MWP [Medieval Warm Period] as "mysterious" and "very incoherent" (his latest pronouncement to me) based on the available data. Of course he and other members of the MBH [Mann Bradley Hughes] camp have a fundamental dislike for the very concept of the MWP, so I tend to view their evaluations as starting out from a somewhat biased perspective, i.e. the cup is not only "half-empty"; it is demonstrably "broken". I come more from the "cup half-full" camp when it comes to the MWP, maybe yes, maybe no, but it is too early to say what it is. Being a natural skeptic, I guess you might lean more towards the MBH camp, which is fine as long as one is honest and open about evaluating the evidence (I have my doubts about the MBH camp). We can always politely(?) disagree given the same admittedly equivocal evidence. I should say that Jan should at least be made aware of this reanalysis of his data. Admittedly, all of the Schweingruber data are in the public domain I believe, so that should not be an issue with those data. I just don't want to get into an open critique of the Esper data because it would just add fuel to the MBH attack squad. They tend to work in their own somewhat agenda-filled ways. We should also work on this stuff on our own, but I do not think that we have an agenda per se, other than trying to objectively understand what is going on.
Cheers,
Ed

From: Keith Briffa [CRU]
To: Edward Cook [Columbia University]
April 29, 2003
Subject: Re: Review- confidential
Thanks Ed
Can I just say that I am not in the MBH [Mann Bradley Hughes] camp—if that be characterized by an unshakable "belief" one way or the other , regarding the absolute magnitude of the global MWP [Medieval Warm Period]. I certainly believe the " medieval" period was warmer than the 18th century—the equivalence of the warmth in the post 1900 period, and the post 1980s ,compared to the circa Medieval times is very much still an area for much better resolution. I think that the geographic / seasonal biases and dating/response time issues still cloud the picture of when and how warm the Medieval period was . On present evidence , even with such uncertainties I would still come out favouring the "likely unprecedented recent warmth" opinion—but our motivation is to further explore the degree of certainty in this

belief—based on the realistic interpretation of available data. Point re Jan well taken and I will inform him

From: Keith Briffa [CRU]
To: Michael E. Mann [University of Virginia]; Tom Wigley [University Corporation of Atmospheric Research]; Phil Jones [CRU]; Raymond Bradley [University of Massachusetts, Amherst]
Cc: Jerry Meehl [University Corporation of Atmospheric Research]; Caspar Ammann [University Corporation of Atmospheric Research]
May 20, 2003
Subject: Re: Soon et al. paper
Mike and Tom and others
. . . As Tom W. states , there are uncertainties and "difficulties" with our current knowledge of Hemispheric temperature histories and valid criticisms or shortcomings in much of our work. This is the nature of the beast—and I have been loathe to become embroiled in polarised debates that force too simplistic a presentation of the state of the art or "consensus view". . . . The one additional point I would make that seems to have been overlooked in the discussions up to now , is the invalidity of assuming that the existence of a global Medieval Warm period , even if shown to be as warm as the current climate , somehow negates the possibility of enhanced greenhouse warming. . . . The various papers apparently in production, regardless of their individual emphasis or approaches, will find their way in to the literature and the next IPCC can sift and present their message(s) as it wishes., but in the meantime , why not a simple statement of the shortcomings of the BS paper as they have been listed in these messages and why not in Climate Research? Keith

From: Tom Wigley [University Corporation of Atmospheric Research]
To: Phil Jones [CRU]
Note: Ben Santer [Lawrence Livermore National Laboratory] may have been Cc'd.
October 21, 2004
[Subject: MBH]
Phil,
I have just read the M&M stuff critcizing MBH [Mann Bradley Hughes]. A lot of it seems valid to me. At the very least MBH is a very sloppy piece of work—an opinion I have held for some time. Presumably what you have done with Keith is better?—or is it? I get asked about this a lot. Can you give me a brief heads up? Mike is too deep into this to be helpful.
Tom.

From: Phil Jones [CRU]
To: Tom Wigley [University Corporation of Atmospheric Research]
Cc: Ben Santer [Lawrence Livermore National Laboratory]
October 22, 2004
Subject: Re: MBH
Tom,
. . . A lot of people criticise MBH [Mann Bradley Hughes] and other papers Mike has been involved in, but how many people read them fully—or just read bits like the attached. The attached is a complete distortion of the facts. M&M are completely wrong in virtually everything they say or do. . . . Mike's may have slightly less variability on decadal scales than the others (especially cf Esper et al), but he is using a lot more data than the others. I reckon they are all biased a little to the summer and none are truly annual—I say all this in the Reviews of Geophysics paper ! Bottom line—their is no way the MWP [Medieval Warm Period] (whenever it was) was as warm globally as the last 20 years. There is also no way a whole decade in the LIA [Little Ice Age] period was more than 1 deg C on a global basis cooler than the 1961-90 mean. This is all gut feeling, no science, but years of experience of dealing with global scales and varaibility. Must got to Florence now. Back in Nov 1.
Cheers
Phil

From: Phil Jones [CRU]
To: Kevin Trenberth [University Corporation of Atmospheric Research]; et al.
December 20, 2004
Subject: Re: [Fwd: Re: [Fwd: Re: "Model Mean Climate" for AR4 [IPCC Fourth Assessment Report]]]
. . . I would like to stick with 1961-90. I don't want to change this until 1981-2010 is complete, for 3 reasons : 1) We need 30 years and 81-10 will get all the MSU in nicely, and 2) I will be near retirement !! 3) is one of perception. As climatologists we are often changing base periods and have done for years. I remember getting a number of comments when I changed from 1951-80 to 1961-90. If we go to a more recent one the anomalies will seem less warm—I know this makes no sense scientifically, but it gives the skeptics something to go on about ! If we do the simple way, they will say we aren't doing it properly. . . .

From: Keith Briffa [CRU]
To: Jonathan Overpeck [University of Arizona]
February ??, 2006
[Subject: bullet debate #3]
Third
I suggest this should be[:]
Taken together , the sparse evidence of Southern Hemisphere temperatures prior to the period of instrumental records indicates that overall warming has occurred during the last 350 years, but the even fewer longer regional records indicate earlier periods that are as warm, or warmer than, 20th century means.
. . . Peck, you have to consider that since the TAR [IPCC Third Assessment Report] , there has been a lot of argument re "hockey stick" and the real independence of the inputs to most subsequent analyses is minimal. True, there have been many different techniques used to aggregate and scale data—but the efficacy of these is still far from established. We should be careful not to push the conclusions beyond what we can securely justify—and this is not much other than a confirmation of the general conclusions of the TAR . We must resist being pushed to present the results such that we will be accused of bias—hence no need to attack Moberg . Just need to show the "most likely"course of temperatures over the last 1300 years—which we do well I think. Strong confirmation of TAR is a good result, given that we discuss uncertainty and base it on more data. Let us not try to over egg the pudding. For what it worth , the above comments are my (honestly long considered) views—and I would not be happy to go further . Of course this discussion now needs to go to the wider Chapter authorship, but do not let Susan [Solomon of NOAA] (or Mike [Michael Mann]) push you (us) beyond where we know is right.

From: Jonathan Overpeck [University of Arizona]
To: Keith Briffa [CRU]
September 13, 2006
. . . I think the second sentence could be more controversial—I don't think our team feels it is valid to say, as they did in TAR [IPCC Third Assessment Report], that "It is also likely that, in the Northern Hemisphere,... 1998 was the warmest year" in the last 1000 years. But, it you think about it for a while, Keith has come up with a clever 2nd sentence (when you insert "Northern Hemisphere" language as I suggest below). At first, my reaction was leave it out, but it grows on you, especially if you acknowledge that many readers will want more explicit prose on the 1998 (2005) issue. . . .

From: David Rind [NASA Goddard Institute for Space Studies]
To: Jonathan Overpeck [University of Arizona]
Cc: Keith Briffa [CRU]; et al.
September 13, 2006
Now getting back to the resolution issue: given what we know about the ability to reconstruct global or NH temperatures in the past—could we really in good conscience say we have the precision from tree rings and the very sparse other data to make any definitive statement of this nature (let alone accuracy)? While I appreciate the cleverness of the second sentence, the problem is everybody will recognize that we are 'being clever'—at what point does one come out looking aggressively defensive? I agree that leaving the first sentence as the only sentence suggests that one is somehow doubting the significance of the recent warm years, which is probably not something we want to do.

From: Jonathan Overpeck [University of Arizona]
To: Keith Briffa [CRU]
Cc: Eystein Jansen [Bjerknes Centre for Climate Research]
January 5, 2005
Subject: Fwd: Re: the Arctic paper and IPCC
. . . I'm still not convinced about the AO recon [Arctic Oscillation reconstruction], and am worried about the late 20th century "coolness" in the proxy recon that's not in the instrumental, but it's a nice piece of work in any case. . . .

From: David Parker [UK Met Office]
To: Neil Plummer [Bureau of Meteorology, Australia]
January 5, 2005
Neil
There is a preference in the atmospheric observations chapter of IPCC AR4 [IPCC Fourth Assessment Report] to stay with the 1961-1990 normals. This is partly because a change of normals confuses users, e.g. anomalies will seem less positive than before if we change to newer normals, so the impression of global warming will be muted. . . .

From: David Rind [NASA Goddard Institute for Space Studies]
To: Keith Briffa [CRU]
January 10, 2005
. . . Well, yes and no. If the mismatch between suggested forcing, model sensitivity, and suggested response for the LIA suggests the forcing is overestimated (in particular the solar forcing), then it makes an earlier

warm period less likely, with little implication for future warming. If it suggests climate sensitivity is really much lower, then it says nothing about the earlier warm period (could still have been driven by solar forcing), but suggests future warming is overestimated. If however it implies the reconstructions are underestimating past climate changes, then it suggests the earlier warm period may well have been warmer than indicated (driven by variability, if nothing else) while suggesting future climate changes will be large. This is the essence of the problem. David

From: Phil Jones [CRU]
To: John Christy [University of Alabama, Huntsville]
July 5, 2005
Subject: This and that
John,
There has been some email traffic in the last few days to a week—quite a bit really, only a small part about MSU. The main part has been one of your House subcommittees wanting Mike Mann and others and IPCC to respond on how they produced their reconstructions and how IPCC produced their report. In case you want to look at this see later in the email ! Also this load of rubbish ! This is from an Australian at BMRC [Bureau of Meteorology Research Centre] (not Neville Nicholls). It began from the attached article. What an idiot. The scientific community would come down on me in no uncertain terms if I said the world had cooled from 1998. OK it has but it is only 7 years of data and it isn't statistically significant.

. . . The Hadley Centre are working on the day/night issue with sondes, but there are a lot of problems as there are very few sites in the tropics with both and where both can be distinguished. My own view if that the sondes are overdoing the cooling wrt MSU4 in the lower stratosphere, and some of this likely (IPCC definition) affects the upper troposphere as well. Sondes are a mess and the fact you get agreement with some of them is miraculous. Have you looked at individual sondes, rather than averages—particularly tropical ones? LKS is good, but the RATPAC update less so.

. . . What will be interesting is to see how IPCC pans out, as we've been told we can't use any article that hasn't been submitted by May 31. This date isn't binding, but Aug 12 is a little more as this is when we must submit our next draft—the one everybody will be able to get access to and comment upon. The science isn't going to stop from now until AR4 [IPCC Fourth Assessment Report] comes out in early 2007, so we are going to have to add in relevant new and important papers. I hope it is up to us to decide what is important and new. So, unless you get something to me soon, it won't be in this version. It shouldn't matter though, as it will be ridiculous to keep later drafts without it. We will be open to criticism though with what we do add in subsequent drafts. Someone is going to check the final version and the Aug 12 draft. This is partly why I've sent you the rest of this email. IPCC, me and whoever will get accused of being political, whatever we do. As you know, I'm not political. If anything, I would like to see the climate change happen, so the science could be proved right, regardless of the consequences. This isn't being political, it is being selfish.

Cheers
Phil

From: Phil Jones [CRU]
To: Neville Nichols [Bureau of Meteorology, Australia]
July 6, 2005
Subject: Fwd: Misc
Neville,

Here's an email from John, with the trend from his latest version in. Also has trends for RATPAC and HadAT2. If you can stress in your talks that it is more likely the sondes are wrong—at least as a group. Some may be OK individually. The tropical ones are the key, but it is these that least is know about except for a few regions. The sondes clearly show too much cooling in the stratosphere (when compared to MSU4), and I reckon this must also affect their upper troposphere trends as well. So, John may be putting too much faith in them wrt agreement with UAH. Happy for you to use the figure, if you don't pass on to anyone else. Watch out for Science though and the Mears/Wentz paper if it ever comes out. Also, do point out that looking at surface trends from 1998 isn't very clever.

Cheers
Phil

From: Neville Nichols [Bureau of Meteorology, Australia]
To: Phil Jones [CRU]
July 6, 2005
[Subject: RE: Misc]
. . . I thought Mike Mann's draft response was pretty good—I had expected something more vigorous, but I think he has got the "tone" pretty right. Do you expect to get a call from Congress?

Neville Nicholls

From: Phil Jones [CRU]
To: Neville Nichols [Bureau of Meteorology, Australia]
July 6th, 2005
Subject: RE: Misc
Neville,
Mike's response could do with a little work, but as you say he's got the tone almost dead on. I hope I don't get a call from congress ! I'm hoping that no-one there realizes I have a US DoE grant and have had this (with Tom W.) for the last 25 years. I'll send on one other email received for interest.
Cheers
Phil

From: Mike MacCracken [Climate Institute]
To: Phil Jones [CRU]; Chris Folland [UK Met Office]
Cc: John Holdren; Rosina Bierbaum
January 3, 2009
Subject: Temperatures in 2009
Dear Phil and Chris—
. . . In any case, if the sulfate hypothesis is right, then your prediction of warming might end up being wrong. I think we have been too readily explaining the slow changes over past decade as a result of variability—that explanation is wearing thin. I would just suggest, as a backup to your prediction, that you also do some checking on the sulfate issue, just so you might have a quantified explanation in case the [warming] prediction is wrong. Otherwise, the Skeptics will be all over us—the world is really cooling, the models are no good, etc. And all this just as the US is about ready to get serious on the issue. We all, and you all in particular, need to be prepared.
Best, Mike MacCracken

From: Tim Johns [UK Met Office]
To: Chris Folland [CRU]
Cc: Doug Smith [UK Met Office]
January 5, 2009
. . . The impact of the two alternative SO2 emissions trajectories is quite marked though in terms of global temperature response in the first few decades of the 21st C (at least in our HadGEM2-AO simulations, reflecting actual aerosol forcings in that model plus some divergence in GHG forcing). Ironically, the E1-IMAGE scenario runs, although

much cooler in the long term of course, are considerably warmer than
A1B-AR4 for several decades! Also—relevant to your statement—
A1B-AR4 runs show potential for a distinct lack of warming in the
early 21st C, which I'm sure skeptics would love to see replicated in
the real world... (See the attached plot for illustration but please don't
circulate this any further as these are results in progress, not yet shared
with other ENSEMBLES partners let alone published). We think the
different short term warming responses are largely attributable to the
different SO2 emissions trajectories. . . .

From: Phil Jones [CRU]
To: Tim Johns [UK Met Office]; Chris Folland [UK Met Office]
Cc: Doug Smith [UK Met Office]
January 5, 2009
Subject: Re: FW: Temperatures in 2009
Tim, Chris,
I hope you're not right about the lack of warming lasting till about 2020.
I'd rather hoped to see the earlier Met Office press release with Doug's
paper that said something like—half the years to 2014 would exceed
the warmest year currently on record, 1998! Still a way to go before
2014. I seem to be getting an email a week from skeptics saying where's
the warming gone. I know the warming is on the decadal scale, but it
would be nice to wear their smug grins away. Chris—I presume the
Met Office continually monitor the weather forecasts. Maybe because
I'm in my 50s, but the language used in the forecasts seems a bit over
the top re the cold. Where I've been for the last 20 days (in Norfolk)
it doesn't seem to have been as cold as the forecasts. . . .

From: Kevin Trenberth [University Corporation of Atmospheric
Research]
To: Michael Mann [Penn State University]
Cc: Stephen Schneider [Stanford University]; Myles Allen [University
of Oxford]; Peter Stott [UK Met Office]; Phil Jones [CRU]; Ben Santer
[Lawrence Livermore National Laboratory]; Tom Wigley [University
Corporation of Atmospheric Research]; Thomas R Karl [NOAA];
Gavin Schmidt [NASA Goddard Institute for Space Studies]; James
Hansen [NASA Goddard Institute for Space Studies]; Michael Oppen-
heimer [Princeton University]
October 12, 2009
Subject: Re: BBC U-turn on climate
Hi all. Well I have my own article on where the heck is global warming?
We are asking that here in Boulder where we have broken records the

past two days for the coldest days on record. We had 4 inches of snow. The high the last 2 days was below 30F and the normal is 69F, and it smashed the previous records for these days by 10F. The low was about 18F and also a record low, well below the previous record low. This is January weather (see the Rockies baseball playoff game was canceled on saturday and then played last night in below freezing weather). The fact is that we can't account for the lack of warming at the moment and it is a travesty that we can't. The CERES data published in the August BAMS 09 supplement on 2008 shows there should be even more warming: but the data are surely wrong. Our observing system is inadequate...

From: Tom Wigley [University Corporation of Atmospheric Research]
To: Phil Jones [CRU]
November 6, 2009
Subject: LAND vs OCEAN
We probably need to say more about this. Land warming since 1980 has been twice the ocean warming—and skeptics might claim that this proves that urban warming is real and important. See attached note. Comments?
Tom

From: Michael E. Mann [University of Virginia]
To: Keith Briffa [CRU]; Tom Wigley [University Corporation of Atmospheric Research]; Phil Jones [CRU]; Raymond Bradley [University of Massachusetts, Amherst]
May 16, 2003
[Subject: Soon et al. paper]
Tom,
Thanks for your response, which I will maintain as confidential within the small group of the original recipients (other than Ray whom I've included in as well), given the sensitivity of some of the comments made. . . . In my view, it is the responsibility of our entire community to fight this intentional disinformation campaign, which represents an affront to everything we do and believe in. I'm doing everything I can to do so, but I can't do it alone—and if I'm left to, we'll lose this battle, mike

From: Michael E. Mann [University of Virginia]
To: Phil Jones [CRU]; Raymond Bradley [University of Massachusetts, Amherst]; Tom Wigley [University Corporation of Atmospheric

Research]; Tom Crowley [Duke University]; Keith Briffa [CRU]; Kevin Trenberth [University Corporation of Atmospheric Research]; Michael Oppenheimer [Princeton University]; Jonathan Overpeck [University of Arizona]

Cc: Scott Rutherford [University of Rhode Island]

June 3, 2003

[Subject: Prospective Eos piece?]

Dear Colleagues,

. . . Phil, Ray, and Peck have already indicated tentative interest in being co-authors. I'm sending this to the rest of you (Tom C, Keith, Tom W, Kevin) in the hopes of broadening the list of co-authors. I strongly believe that a piece of this sort co-authored by 9 or so prominent members of the climate research community (with background and/or interest in paleoclimate) will go a long way ih helping to counter these attacks, which are being used, in turn, to launch attacks against IPCC....

From: Michael E. Mann [University of Virginia]

To: Phil Jones [CRU]; et al.

June 4, 2003

Subject: Re: Prospective Eos piece?

Phil and I have recently submitted a paper using about a dozen NH [Northern Hemisphere] records that fit this category, and many of which are available nearly 2K [2 thousand years] back—I think that trying to adopt a timeframe of 2K, rather than the usual 1K, addresses a good earlier point that Peck [Jonathan Overpeck—University of Arizona] made w/ regard to the memo, that it would be nice to try to "contain" the putative "MWP" [Medieval Warm Period], even if we don't yet have a hemispheric mean reconstruction available that far back [Phil and I have one in review—not sure it is kosher to show that yet though—I've put in an inquiry to Judy Jacobs at AGU about this]. . . .

From: Phil Jones [CRU]

To: Michael E. Mann [University of Virginia]

June 4, 2003

[Subject: Prospective Eos piece?]

. . . EOS would get to most fellow scientists. As I said to you the other day, it is amazing how far and wide the SB pieces have managed to percolate. When it comes out I would hope that AGU/EOS 'publicity machine' will shout the message from rooftops everywhere. As many of us need to be available when it comes out. There is still no firm news on what Climate Research will do, although they will likely

have two editors for potentially controversial papers, and the editors will consult when papers get different reviews. All standard practice I'd have thought. At present the editors get no guidance whatsoever. It would seem that if they don't know what standard practice is then they shouldn't be doing the job !
Cheers
Phil

From: Phil Jones [CRU]
To: Janice Lough [Australian Institute of Marine Science]
August 6th, 2004
Subject: Re: liked the paper
. . . PS Do you want to get involved in IPCC this time? I'm the CLA [Coordinating Lead Author] of the atmospheric obs. [observations] chapter with Kevin Trenberth and we'll be looking for Contributing Authors to help the Lead Authors we have. Paleo[climatology] is in a different section this time led by Peck and Eystein Janssen. Keith is a lead author as well.

From: Phil Jones [CRU]
To: Michael E. Mann [Penn State University]
May 19, 2009
[Subject: nomination: materials needed!]
. . . Apart from my meetings I have skeptics on my back—still, can't seem to get rid of them. Also the new UK climate scenarios are giving govt ministers the jitters as they don't want to appear stupid when they introduce them (late June?). . . .

From: Narsimha D. Rao [Stanford University]
To: Stephen H. Schneider [Stanford University]
October 11, 2009
Subject: BBC U-turn on climate
Steve, You may be aware of this already. Paul Hudson, BBCs reporter on climate change, on Friday wrote that theres been no warming since 1998, and that pacific oscillations will force cooling for the next 20-30 years. It is not outrageously biased in presentation as are other skeptics views. . . . BBC has significant influence on public opinion outside the US. Do you think this merits an op-ed response in the BBC from a scientist?

From: Michael E. Mann [Penn State University]
To: Stephen H. Schneider [Stanford University]

Cc: Myles Allen [University of Oxford]; Peter Stott [UK Met Office]; Phil Jones [CRU]; Ben Santer [Lawrence Livermore National Laboratory]; Tom Wigley [University Corporation of Atmospheric Research]; Thomas R Karl [NOAA]; Gavin Schmidt [NASA Goddard Institute for Space Studies]; James Hansen [NASA Goddard Institute for Space Studies]; Kevin Trenberth [University Corporation of Atmospheric Research]; Michael Oppenheimer [Princeton University]
October 12, 2009
Subject: Re: BBC U-turn on climate
extremely disappointing to see something like this appear on BBC. its particularly odd, since climate is usually Richard Black's beat at BBC (and he does a great job). from what I can tell, this guy was formerly a weather person at the Met Office. We may do something about this on RealClimate [website], but meanwhile it might be appropriate for the Met Office [UK's National Weather Service] to have a say about this, I might ask Richard Black [BBC environment correspondent] what's up here?

From: Phil Jones [CRU]
To: Gavin Schmidt [NASA Goddard Institute for Space Studies]; Michael E. Mann [Penn State University]; Andy Revkin [New York Times]
October 27, 2009
[Subject: The web page is up about the Yamal tree-ring chronology]
Gavin, Mike, Andy,
It has taken Keith longer than he would have liked, but it is up. There is a lot to read and understand. It is structured for different levels. The link goes to the top level. There is more detail below this and then there are the data below that. . . . I'll let you make up you own minds! It seems to me as though McIntyre cherry picked for effect. There is an additional part that shows how many series from Ch 6 of AR4 [IPCC Fourth Assessment Report] used Yamal—most didn't!

From: Michael E. Mann [Penn State University]
To: Phil Jones [CRU]
Note: Gavin Schmidt [NASA Goddard Institute for Space Studies] may have been cc'd.
October 27, 2009
[Subject: The web page is up about the Yamal tree-ring chronology]
thanks Phil,
Perhaps we'll do a simple update to the Yamal post, e.g. linking Keith/s new page—Gavin t? As to the issues of robustness, particularly w.r.t.

inclusion of the Yamal series, we actually emphasized that (including the Osborn and Briffa '06 sensitivity test) in our original post! As we all know, this isn't about truth at all, its about plausibly deniable accusations,

m

From: Michael E. Mann [Penn State University]
To: Phil Jones [CRU]
Note: Gavin Schmidt [NASA Goddard Institute for Space Studies] may have been cc'd.
October 27, 2009
[Subject: The web page is up about the Yamal tree-ring chronology]
Hi Phil,

Thanks—we know that. The point is simply that if we want to talk about about a meaningful "2009" anomaly, every additional month that is available from which to calculate an annual mean makes the number more credible. We already have this for GISTEMP, but have been awaiting HadCRU to be able to do a more decisive update of the status of the disingenuous "globe is cooling" contrarian talking point, mike

p.s. be a bit careful about what information you send to Andy [Revkin with the New York Times] and what emails you copy him in on. He's not as predictable as we'd like

"HARRY READ ME" FILE

Among CRU's exposed documents is the so-called "HARRY_READ_ME" file, which served as a detailed note keeping file from 2006 through 2009 for CRU researcher and programmer Ian "Harry" Harris. As he worked to update and modify CRU TS2.1 to create the new CRU TS3.1dataset, the HARRY_READ_ME.txt details Harris's frustration with the dubious nature of CRU's meteorological datasets. As demonstrated through a handful of excerpts below, the 93,000-word HARRY_READ_ME file may raise several serious questions as to the reliability and integrity of CRU's data compilation and quality assurance protocols.

I am very sorry to report that the rest of the databases seem to be in nearly as poor a state as Australia was. There are hundreds if not thousands of pairs of dummy stations, one with no WMO and one with, usually overlapping and with the same station name and very similar coordinates. I know it could be old and new stations, but why such large overlaps if that's the case? Aarrggghhh! There truly is no end in sight.

One thing that's unsettling is that many of the assigned WMo codes for Canadian stations do not return any hits with a web search. Usually the country's met office, or at least the Weather Underground, show up—but for these stations, nothing at all. Makes me wonder if these are long-discontinued, or were even invented somewhere other than Canada!

OH F**K THIS. It's Sunday evening, I've worked all weekend, and just when I thought it was done I'm hitting yet another problem that's based on the hopeless state of our databases. There is no uniform data integrity, it's just a catalogue of issues that continues to grow as they're found.

Here, the expected 1990-2003 period is MISSING—so the correlations aren't so hot! Yet the WMO codes and station names /locations are identical (or close). What the hell is supposed to happen here? Oh yeah—there is no 'supposed', I can make it up. So I have :-)

You can't imagine what this has cost me—to actually allow the operator to assign false WMO codes!! But what else is there in such situations? Especially when dealing with a 'Master' database of dubious provenance (which, er, they all are and always will be).

False codes will be obtained by multiplying the legitimate code (5 digits) by 100, then adding 1 at a time until a number is found with no matches in the database. THIS IS NOT PERFECT but as there is no central repository for WMO codes—especially made-up ones—we'll have to chance duplicating one that's present in one of the other databases. In any case, anyone comparing WMO codes between databases—something

I've studiously avoided doing except for tmin/tmax where I had to—will be treating the false codes with suspicion anyway. Hopefully.

Of course, option 3 cannot be offered for CLIMAT bulletins, there being no metadata with which to form a new station.

This still meant an awful lot of encounters with naughty Master stations, when really I suspect nobody else gives a hoot about. So with a somewhat cynical shrug, I added the nuclear option— to match every WMO possible, and turn the rest into new stations (er, CLIMAT excepted). In other words, what CRU usually do. It will allow bad databases to pass unnoticed, and good databases to become bad, but I really don't think people care enough to fix 'em, and it's the main reason the project is nearly a year late.

This whole project is SUCH A MESS. No wonder I needed therapy!!

So.. we don't have the coefficients files (just .eps plots of something). But what are all those monthly files? DON'T KNOW, UNDOCUMENTED. Wherever I look, there are data files, no info about what they are other than their names. And that's useless.. take the above example, the filenames in the _mon and _ann directories are identical, but the contents are not. And the only difference is that one directory is apparently 'monthly' and the other 'annual'—yet both contain monthly files.

I find that they are broadly similar, except the normals lines (which both start with '6190') are very different. I was expecting that maybe the latter contained 94-00 normals, what I wasn't expecting was that are in % x10 not %! Unbelievable—even here the conventions have not been followed. It's botch after botch after botch. Modified the conversion program to process either kind of normals line.

The biggest immediate problem was the loss of an hour's edits to the program, when the network died.. no explanations from anyone, I hope it's not a return to last year's troubles.

(some weeks later)

well, it compiles OK, and even runs enthusiastically. However there are loads of bugs that I now have to fix. Eeeeek. Timesrunningouttimesrunningout.

(even later)

Getting there.. still ironing out glitches and poor programming.

25. Wahey! It's halfway through April and I'm still working on it. This surely is the worst project I've ever attempted. Eeeek.

So the 'duplicated' figure is slightly lower.. but what's this error with the '.ann' file?! Never seen before. Oh GOD if I could start this project again and actually argue the case for junking the inherited program suite!!

Wrote 'makedtr.for' to tackle the thorny problem of the tmin and tmax databases not being kept in step. Sounds familiar, if worrying. am I the first person to attempt to get the CRU databases in working order?!! The program pulls no punches.

Back to the gridding. I am seriously worried that our flagship gridded data product is produced by Delaunay triangulation—apparently linear as well. As far as I can see, this renders the station counts totally meaningless. It also means that we cannot say exactly how the gridded data is arrived at from a statistical perspective—since we're using an off-the-shelf product that isn't documented sufficiently to say that. Why this wasn't coded up in Fortran I don't know—time pressures perhaps?

Was too much effort expended on homogenisation, that there wasn't enough time to write a gridding procedure? Of course, it's too late for me to fix it too. Meh.

Now looking at the dates.. something bad has happened, hasn't it. COBAR AIRPORT AWS cannot start in 1962, it didn't open until 1993! Looking at the data—the COBAR station 1962-2004 seems to be an exact copy of the COBAR AIRPORT AWS station 1962-2004, except that the latter has more missing values. Now, COBAR AIRPORT AWS has 15 months of missing value codes beginning Oct 1993.. coincidence?

I am seriously close to giving up, again. The history of this is so complex that I can't get far enough into it before by head hurts and I have to stop. Each parameter has a tortuous history of manual and semi-automated interventions that I simply cannot just go back to early versions and run the update prog. I could be throwing away all kinds of corrections—to lat/lons, to WMOs (yes!), and more.

So what the hell can I do about all these duplicate stations? Well, how about fixdupes.for? That would be perfect—except that I never finished it, I was diverted off to fight some other fire. Aarrgghhh.

I—need—a—database—cleaner.

What about the ones I used for the CRUTEM3 work with Phil Brohan? Can't find the bugger!! Looked everywhere, Matlab scripts aplenty but not the one that produced the plots I used in my CRU presentation in 2005. Oh, F**K IT. Sorry. I will have to WRITE a program to find potential duplicates. It can show me pairs of headers, and correlations between the data, and I can say 'yay' or 'nay'. There is the finddupes.for program, though I think the comment for *this* program sums it up nicely:

` program postprocdupes2

c Further post-processing of the duplicates file—just to show how crap the

c program that produced it was! Well—not so much that but that once it was

c running, it took 2 days to finish so I couldn't really reset it to improve

c things. Anyway, *this* version does the following useful stuff:

c (1) Removes and squirrels away all segments where dates don't match;

c (2) Marks segments >5 where dates don't match;

c (3) Groups segments from the same pair of stations;

c (4) Sorts based on total segment length for each station pair'

You see how messy it gets when you actually examine the problem?

Well, dtr2cld is not the world's most complicated program. Wheras cloudreg is, and I immediately found a mistake! Scanning forward to 1951 was done with a loop that, for completely unfathomable reasons, didn't include months! So we read 50 grids instead of 600!!! That may have had something to do with it. I also noticed, as I was correcting THAT, that I reopened the DTR and CLD data files when I should have been opening the bloody station files!! I can only assume that I was being interrupted continually when I was writing this thing. Running with those bits fixed improved matters somewhat, though now there's a problem in that one 5-degree band (10S to 5S) has no stations! This will be due to low station counts in that region, plus removal of duplicate values.

WHY?

THE FORCES ARE STILL OUT THERE and, with or without President Obama, they have the commitment and resources to achieve their Crown Jewel.

As I said in the beginning...Why?

When the United Nations is totally refuted...

When Al Gore is totally discredited...

When man-made catastrophic global warming is totally debunked...

When passing global warming cap and trade is totally futile...

Why is this book necessary?

Now you know.

ENDNOTES

INTRODUCTION: THE HOAX DEBUNKED: DON'T FEEL TOO SORRY FOR AL GORE

1   Senator James Inhofe, "Catastrophic Global Warming Alarmism Not Based on Objective Science," Senate Floor Speech, July 29, 2003 http://epw.senate.gov/public/index. cfm?FuseAction=Minority.PressReleases&ContentRecord_id=037e5dea-9c4e-4b92-a416-4a6bf6cfd7ef&Region_id=&Issue_id.

2   Al Gore, Op-Ed, "We Can't Wish Away Climate Change," New York Times, February 27, 2010, http://www.nytimes.com/2010/02/28/opinion/28gore.html?pagewanted=all.

3   Christopher Booker, Climate change: this is the worst scientific scandal of our generation, The Telegraph, November 28, 2009
http://www.telegraph.co.uk/comment/columnists/christopherbooker/6679082/Climate-change-this-is-the-worst-scientific-scandal-of-our-generation.html

4   Clive Crook, Death of the IPCC, the Atlantic, February 9, 2010 http://www.theatlantic.

com/business/archive/2010/02/death-of-the-ipcc/35618/

5   Editorial, Time for a change in climate research, Financial Times, August 31, 2010 http://www.ft.com/intl/cms/s/0/362ddb00-b534-11df-9af8-00144feabdc0.html#axzz1cKkXJQcJ

6   Leslie Kaufman, Among Weathercasters, Doubt on Warming, New York Times, March 29, 2010 http://www.nytimes.com/2010/03/30/science/earth/30warming.html?src=me

7   Editorial, Uncertain Science, Newsweek, May 27, 2010 http://www.thedailybeast.com/newsweek/2010/05/28/uncertain-science.html

8   Bryan Walsh, Himalayan Melting: How a Climate Panel Got It Wrong, Time Magazine, January 21, 2010 http://www.time.com/time/health/article/0,8599,1955405,00.html#ixzz1aWnXy6cw

9   Another Ice Age?, Time Magazine, June 24, 1974 http://www.time.com/time/magazine/article/0,9171,944914,00.html

10 Be Worried, Be Very Worried," Time Magazine, April 3, 2006, http://www.time.com/time/covers/0,16641,20060403,00.html.

11 Senator Barack Obama, Text of Democrat Barack Obama's Prepared Remarks for Rally in St. Paul, Minnesota, "Defining Moment for Our Nation," Associated Press July 3, 2008, http://www.breitbart.com/article.php?id=D912VD200.

12 "New Video: Obama Vows Electricity Rates Would 'Necessarily Skyrocket' Under His Plan, San Francisco Chronicle, January 2008, Posted on Breitbart.tv, November 3, 2008, http://tv.breitbart.com/obama-vows-electricity-rates-would-necessarily-skyrocket-under-his-plan/.

13 Global Warming: The High Cost of The Kyoto Protocol." WEFA, Inc. 1998. http://www.accf.org/media/dynamic/9/media_98.pdf.

14 "Analysis of State-Level Economic Impacts of the McCain-Lieberman Bill," Charles River Associates study, June 2004 http://news.heartland.org/newspaper-article/2004/08/01/new-study-shows-hefty-price-tag-mccain-lieberman-bill

15 Senator Inhofe Press Release," Jackson Confirms EPA Chart Showing No Effect on Climate without China, India," Senate Environment and Public Works Committee, July 7, 2009, http://epw.senate.gov/public/index.cfm?FuseAction=Minority.PressReleases&ContentRecord_id=564ed42f-802a-23ad-4570-3399477b1393&Region_id=&Issue_id=.

16 Wigley, T.M.L. 1998. The Kyoto Protocol: CO2, CH4 and climate implications. Geophysical Research Letters, Vol. 25, No. 13: 2285-88. http://www2.ucar.edu/news/record/effect-kyoto-protocol-global-warming#grl

17 Al Gore, Op-Ed, Climate of Denial, Rolling Stone, June 22, 2011 http://www.rollingstone.com/politics/news/climate-of-denial-20110622

18 The Environmental Protection Agency, Endangerment and Cause or Contribute Findings for Greenhouse Gases under Section 202(a) of the Clean Air Act http://epa.gov/climatechange/endangerment.html

19 John Bryson, Envisioning the Future of Global Energy, University of California–Berkeley Energy Symposium, March 4, 2010, http://www.viddler.com/explore/BERCTV/videos/27/. See also http://epw.senate.gov/public/index.cfm?FuseAction=Minority.PressReleases&ContentRecord_id=47843c26-802a-23ad-466d-b17b60a5fef1&Region_id&Issue_id.

20 Rebecca Wodder, Blog Post,"Turning Endangered Rivers into Success Stories," Huffington Post, June 2, 2010, http://www.huffingtonpost.com/rebecca-wodder/turning-endangered-rivers_b_589556.html.

21 Alan B. Krueger, Testimony before the Senate Committee on Finance, Subcommittee on Energy, Natural Resources, and Infrastructure, September 10, 2009

22 Neil King Jr. and Stephen Power,"Times Tough for Energy Overhaul," Wall Street Journal, December 12, 2008, http://online.wsj.com/article/SB122904040307499791.html.

23 Inhofe Press Release, Inhofe Welcomes EPA Delay of Greenhouse Gas Regulations,

September 15, 2011 http://epw.senate.gov/public/index.cfm?FuseAction=Minority. PressReleases&ContentRecord_id=6d85044c-802a-23ad-439c-a5e26bff6c16&Region_id=&Issue_id=

24   IPCC, 2007: Summary for Policymakers. In: Climate Change 2007: The Physical Science Basis. Contribution of Working Group I to the Fourth Assessment Report of the Intergovernmental Panel on Climate Change [Solomon, S., D. Qin, M. Manning, Z. Chen, M. Marquis, K.B. Averyt, M.Tignor and H.L. Miller (eds.)]. Cambridge University Press, Cambridge, United Kingdom and New York, NY, USA.

25   "24% Consider Al Gore and Expert on Global Warming," Rasmussen, September 8, 2011 http://www.rasmussenreports.com/public_content/politics/current_events/environment_energy/24_consider_al_gore_an_expert_on_global_warming

26   Steven Hayward, "In Denial: The Meltdown of the Climate Campaign," The Weekly Standard, March 15, 2010. http://www.weeklystandard.com/articles/denial

27   John Broder, Gore's New Role: Advocate and Investor, New York Times, November 2, 2009 http://www.nytimes.com/2009/11/03/business/energy-environment/03gore.html

28   Stephen Spruiell, Climate Profiteers, For Gore & Co., green is gold," National Review Online, March 22, 2010,

29   Broder, Gore's New Role: Advocate and Investor, 2009.

30   Ibid

1: WHY I FIGHT

31   Caroline May, Sen. Inhofe turns tables on global warming ambushers – and gets it on tape [Video], Daily Caller, February 18, 2011, http://dailycaller.com/2011/02/18/sen-inhofe-turns-tables-on-global-warming-ambushers-and-gets-it-on-tape-video/

32   John Stanton, "Inhofe is Happy to Stand Apart," Roll Call, November 22, 2010, http://www.rollcall.com/issues/56_50/-200810-1.html.

33   The Daily Oklahoman, Remodeling the House, 10/12/1993

34   Eric Fox, "Bubbling Crude: America's Top 6 Producing States," MSNBC, June 8, 2011, http://www.msnbc.msn.com/id/43085246/ns/business-oil_and_energy/t/bubbling-crude-americas-top-oil-producing-states/

35   U.S. Energy Information Administration, Oklahoma Profile, Data – Shares and Supply http://205.254.135.24/state/state-energy-profiles.cfm?sid=OK

36   Jay F. Marks, "Study Says Oklahoma Economy Got Up to $51B from Oil and Gas Industry in 2009," Daily Oklahoman, June 23, 2011. http://newsok.com/study-says-oklahoma-economy-got-up-to-51b-from-oil-and-gas-industry-in-2009/article/3579502#ixzz1QCQbFmDc

37   Emily Belz, Going green: It's the new red, white and blue, The Hill, August 3, 2007 http://thehill.com/capital-living/23946-going-green-its-the-new-red-white-and-blue

38   Senator James M. Inhofe, "The U. S. Senate Minority Report: More Than 700 International Scientists Dissent over Man-Made Global Warming Claims – Scientists Continue to Debunk 'Consensus,'" version March 23, 2009.

2: "THE MOST DANGEROUS MAN ON THE PLANET": EXPOSING THE HOAX

39   "Letter to William J. Clinton on the Kyoto Protocol," Business Roundtable, May 12, 1998, research.greenpeaceusa.org/?a=download&d=2052.

40   Byrd-Hagel Resolution, S. Res 98, 105th Congress, 1st Session. Introduced June 12, 1997. http://thomas.loc.gov/cgi-bin/bdquery/z?d105:SE00098:#.

41   Global Warming: The High Cost of The Kyoto Protocol." WEFA, Inc. 1998. http://www.
       accf.org/media/dynamic/9/media_98.pdf.

42   Congressional Budget Office, "Shifting the Cost Burden of a Carbon Cap-and-Trade Pro-
       gram." July 2003, http://www.cbo.gov/doc.cfm?index=4401&type=0.

43   J. Thomas Mullen, Testimony before the Senate Committee on Environment and Public
       Works, June 12, 2002 http://epw.senate.gov/107th/Mullen_061202.htm

44   Germany released a statement declaring that the world needs Kyoto because its green-
       house gas reduction targets "are indispensable."   http://epw.senate.gov/pressitem.
       cfm?party=rep&id=220330

45   Summit fails to solve climate dispute, BBC News, June 14, 2001 http://news.bbc.co.uk/2/
       hi/europe/1387667.stm

46   "Speech by Mr. Jacques Chirac, French President, to the VIth Conference of the Parties to
       the United Nations Framework Convention on Climate Change." The Hague. Nov. 20, 2000.
       http://sovereignty.net/center/chirac.html

47   Stephen Castle, "EU sends strong warning to Bush over greenhouse gas emissions," Inde-
       pendent UK, March 19, 2001, http://www.independent.co.uk/news/world/europe/eu-sends-
       strong-warning-to-bushover-greenhouse-gas-emisssions-687997.html. (link not working
       right now). Also found here: http://www.commondreams.org/cgi-bin/print.cgi?file=/head-
       lines01/0319-01.htm

48   Harper's letter dismisses Kyoto as 'socialist scheme', CBC News, January 30, 2007 http://
       www.cbc.ca/news/canada/story/2007/01/30/harper-kyoto.html

49   Editorial, "Cold Shoulder, Inhofe Holds His Own on Global Warming Debate," Oklahoman,
       October 9, 2006, http://newsok.com/article/2952076/.

50   Peter Gwyne, "The Cooling World," Newsweek, April 28, 1975. http://denisdutton.com/
       newsweek_coolingworld.pdf.

51   "Another Ice Age?" Time, June 24, 1974, http://www.time.com/time/magazine/
       article/0,9171,944914,00.html.

52   National Science Board, Science and the Challenges Ahead, 1974 Page 24 http://www.archive.
       org/details/sciencechallenge00nati

53   National Science Board, Patterns and Perspectives in Environmental Science, 1972, http://
       www.archive.org/stream/patternsperspect00nati/patternsperspect00nati_djvu.txt

54   Eduardo A. Cavallo, Anderw Powerll & Oscar Becerra, Estimating the Direct Economic
       Damage of the Earthquake in Haiti, IDB Working Paper Series No. IDB-WP-163, February
       2010, accessed from http://idbdocs.iadb.org/wsdocs/getdocument.aspx?docnum=35074108
       on 10,28,2011

55   Robert Frank, Americans Raise Record $150 Million for Haiti, Wall Street Journal, January
       18, 2010, http://blogs.wsj.com/wealth/2010/01/18/americans-raise-record-150-million-for-
       haiti/

56   Stephen Castle, "EU sends strong warning to Bush over greenhouse gas emissions," Inde-
       pendent UK, March 19, 2001, http://www.independent.co.uk/news/world/europe/eu-sends-
       strong-warning-to-bushover-greenhouse-gas-emisssions-687997.html.

57   United Nations, About US, Overview of the UN Global Compact, www.unglobalcompact.
       org/AboutTheGC

58   United Nations, United Nations Framework Convention on Climate Change, 1992, p.5.
       http://unfccc.int/resource/docs/convkp/conveng.pdf.

59   S. Fred Singer, Testimony before the House Small Business Committee, July 29, 1998.

60   Wigley, T.M.L. The Kyoto Protocol: CO2, CH4 and climate implications. Geophysical
       Research Letters, Vol. 25, No. 13: 2285-88. 1998. http://www2.ucar.edu/news/record/
       effect-kyoto-protocol-global-warming#grl

61   Dr. Richard S. Lindzen, Testimony before the Senate Committee on Environment and Public

Works Committee, May 2, 2001, http://epw.senate.gov/107th/lin_0502.htm.

62 James Hansen, et al., "Global Warming in the 21st Century: An Alternative Scenario," http://www.giss.nasa.gov/research/features/200111_altscenario/.

63 Chris Mooney, "Some Like It Hot," Mother Jones, May/June 2005 Issue, http://motherjones.com/environment/2005/05/some-it-hot.

64 Frederick Seitz, Op-Ed, "A Major Deception on 'Global Warming,'" Wall Street Journal, June 12, 1996.

65 IPCC 95.1996 Climate Change 1995: The Science of Climate Change. Cambridge University Press, UK.

66 William Stevens, "Global Warming Experts Call Human Role Likely, "New York Times, September 11, 1995.

67 John Christy, Op-Ed, "So-Called Global Warming 'Data' Frequently Misleads," The Montgomery Advertiser, February 22, 1998.

68 Ibid

69 Editorial, "The Warming Debate," Washington Post, October 30, 2000.

70 Editorial, "A Sharper Warning on Warming," New York Times, October 28, 2000.

71 Richard Lindzen, "Testimony before the Senate Committee on Environment and Public Works," May 2, 2001. http://epw.senate.gov/107th/lin_0502.htm

72 Ibid

73 Patrick Michaels, "Sound and Fury: the Science and Politics of Global Warming," Cato Institute, 1992

74 "20th Century Climate Not So Hot," Press Release, Harvard-Smithsonian Center for Astrophysics, March 31, 2003, http://www.cfa.harvard.edu/news/archive/pr0310.html

75 Interview, Der Spiegel, No. 41-2004, October 4, 2004, 158.

76 Backgrounder for McIntrye and McKitrick, "Hockey Stick Project," January 27, 2005.

77 McIntyre and McKitrick, "Hockey Sticks, Principal Components and Spurious Significance," Geophysical Research Letters, American Geophysical Union 2005.

78 Antonio Regalado, "In Climate Debate, 'The Hockey Stick' Lead to a Face-Off," Wall Street Journal, February 14, 2005.

79 Ibid

80 David Deming, "Testimony Before the US Senate Committee on Environment and Public Works, December 6, 2006. http://epw.senate.gov/hearing_statements.cfm?id=266543

81 Open Letter from Chris Landsea, January 17, 2005. http://cstpr.colorado.edu/prometheus/archives/science_policy_general/000318chris_landsea_leaves.html

82 Lord Nigel Lawson, Testimony Before the US Senate Committee on Environment and Public Works, December 5, 2005. http://epw.senate.gov/hearing_statements.cfm?id=246944

83 Inhofe Letter to Dr. R.K. Pachauri, December 7, 2005 http://epw.senate.gov/public/index.cfm?FuseAction=PressRoom.PressReleases&ContentRecord_id=B5B423AC-B42C-4879-91CE-348E78AAE64B

84 Alister Doyle, Pachauri rebuts bias charges, Reuters, October 12, 2007 http://www.reuters.com/article/2007/10/13/environment-nobel-pachauri-dc-idUSL128894720071013

85 Michael Crichton, State of Fear, Harper Collins, First Edition 2004, 401-402.

86 Ibid, 245-246

87 Ibid, 295

88 Ibid, 314

89 Jack Williams, Scientists, others trying to make sense of Arctic changes, USA Today, October 29, 2003 http://www.usatoday.com/weather/resources/coldscience/2003-10-29-arctic-searc_x.htm

90 Inhofe, Voinovich Press Release, Inhofe, Voinovich Introduce Clear Skies Legislation, January 25, 2005 http://epw.senate.gov/pressitem.cfm?party=rep&id=230820

3: THE BUILD-UP OF ALARMISM

91 Alessandra Stanley, Sounding the Global Warming Alarm Without Upsetting the Fans, New York Times, Sept. 9, 2007 http://www.nytimes.com/2007/07/09/arts/television/09watc.html?_r=1&oref=slogin&pagewanted=print

92 Laurie David, "Oprah Sees the Light (Bulb)," Huffington Post, November 2, 2005, http://www.huffingtonpost.com/laurie-david/oprah-sees-the-light-bulb_b_9999.html.

93 Sheryl Crow, "Laurie and Sheryl Go to School," Huffington Post, April 19, 2007, http://www.huffingtonpost.com/sheryl-crow/laurie-and-sheryl-go-to-s_b_46320.html.

94 Transcript, "Countdown with Keith Olbermann,' for March 21," MSNBC, http://www.msnbc.msn.com/id/17737423/ns/msnbc_tv-countdown_with_keith_olbermann/t/countdown-keith-olbermann-march/.

95 Katie Couric, "Gore Warms Up to Hollywood," CBS News, February 26, 2007, http://www.cbsnews.com/8301-500803_162-2518971-500803.html.

96 "Al Gore on Oprah—What to Do about Global Warming," Everything Oprah: The Latest Oprah Winfrey Show Highlights, November 28, 2008, http://www.everythingoprah.com/2008/11/al-gore-on-oprah-what-to-do-about-global-warming.html.

97 Richard Lindzen, Op-Ed, "There Is No 'Consensus' On Global Warming," The Wall Street Journal, June 26, 2006, http://epw.senate.gov/public/index.cfm?FuseAction=Minority.Blogs&ContentRecord_id=5e3e21f5-1b4c-4208-912d-87d0a5ba342b&Issue_id

98 Earth to America, TBS, November 20, 2005.

99 Al Gore, with Billy West, An Inconvenient Truth (film documentary), directed by Davis Guggenheim, produced by Lawrence Bender, Scott Z. Burns, and Laurie David; distributed by Paramount Classics, 2006. Based on the teleplay by Al Gore.

100 "DiCaprio Sheds light on '11th Hour,'" The Hollywood Reporter, May 20, 2007, http://www.hollywoodreporter.com/news/dicaprio-sheds-light-11th-hour-136753.

101 "Leonardo DiCaprio talks to USA WEEKEND," USA Weekend, September 20, 2007.

102 Laurie David, To Our Three Daughters, Huffington Post, May 12, 2007. http://www.huffingtonpost.com/laurie-david/to-our-three-daughters_b_48293.html

103 Darragh Johnson, "Climate Change Scenarios Scare, and Motivate, Kids," Washington Post, April 16, 2007, http://www.washingtonpost.com/wp-dyn/content/article/2007/04/15/AR2007041501164_pf.html .

104 "Prospects of Another Glacial Period: Geologists Think the World May be Frozen Up Again," New York Times, February 24, 1895, p. 6.

105 Glacial Era Coming: Prof. Schmidt Warns Us of an Enroaching Ice Age," New York Times, October 7, 1912, p. 1

106 "Human race will have to fight for its existence against the cold," Los Angeles Times, October 7, 1912

107 "Ice Age Coming Here," Washington Post, August 10, 1923.

108 America in Longest Warm Spell Since 1776," New York Times, March 27, 1933

109 "Scientists Says Arctic Ice Will Wipe Out Canada" Chicago Tribune, August 9, 1923

110 "Present climate change will result in mass deaths by starvation and probably in anarchy and violence" New York Times, December 29, 1974

111 "Scientist ponder why World's Climate is changing; a major cooling is considered to be inevitable" New York Times, May 21, 1975

112 "Be Worried, Be Very Worried," Time magazine, April 3, 2006, http://www.time.com/time/covers/0,16641,20060403,00.html.

113 Jerry Adler, "From Newsweek: Remember Global Cooling?" Daily Beast, October 22, 2006. http://www.thedailybeast.com/newsweek/2006/10/22/remember-global-cooling.html.

114 Peter Gwyne, "The Cooling World," Newsweek, April 28, 1975, 64.

115 Jerry Adler, Newsweek, October 22, 2006

116 Ibid

117 Ibid

118 Ibid

119 Editorial, "Doubting Inhofe," New York Times, October 12, 2006, http://www.nytimes.com/2006/10/12/opinion/12thu2.html.

120 American Morning, CNN Transcript, October 3, 2006.

121 Associated Press, "Scientists Give Two Thumbs Up to Gore's Move on Global Warming," USA Today, June 27, 2006, http://www.usatoday.com/weather/news/2006-06-27-inconvenient-truth-reviews_x.htm.

122 National Academy of Sciences, Temperature Reconstruction for the Last 2,000 Years, June 2006.

123 Seth Borenstein, "Scientists OK Gore's Movie for Accuracy," Associated Press, June 27, 2006, http://www.washingtonpost.com/wp-dyn/content/article/2006/06/27/AR2006062700780_pf.html.

124 Bill Blakemore, "Schwarzenator' vs. Bush: Global Warming Debate Heats Up," ABC News, August 30, 2006. http://abcnews.go.com/US/GlobalWarming/story?id=2374968&page=1

125 Mark, Finkelstein, ABC Trolls for Global Warming Stories, Newsbusters, August 2, 2006.

126 Dave Shiflett, "Brokaw Warns of Melting Glaciers, Greenhouse Gases; TV Review, Bloomberg, July 14, 2006, quoted in "Brokaw's Global Warming Show = Less Dissent Than 'North Korean Political Rally' -- Bloomberg TV Review Says," U.S. Senate Committee on Environment & Public Works website, http://epw.senate.gov/fact.cfm?party=rep&id=258659.

127 Roger Pielke Sr., TV Special "Global Warming" to Air July 16, 2006, Climate Science: Roger Pielke Sr, July 7, 2006, http://pielkeclimatesci.wordpress.com/2006/07/07/nbcdiscovery-channel-show/.

128 "Rewriting The Science," 60 Minutes, CBS News, October 8, 2009, http://www.cbsnews.com/stories/2006/03/17/60minutes/main1415985.shtml.

129 James Hansen, "Can We Defuse the Global Warming Time Bomb, Appendix: Climate forcing scenarios," August 1, 2003, http://naturalscience.com/ns/articles/01-16/ns_jeh6.html.

130 Fox News Reporting, "The Heat Is On: The Case of Global Warming," Fox News, November 11, 2005, http://www.foxnews.com/on-air/fox-news-reporting/2005/11/11/heat-case-global-warming-0.

131 Matea Gold, "Fox News displays a green side," Los Angeles Times, November 12, 2005, http://articles.latimes.com/2005/nov/12/entertainment/et-david12.

132 Marc Morano, CBS News Seeks 'Hip' Environmental Reporter, No 'Knowledge of Enviro Beat' Necessary, Inhofe EPW Press Blog, November 30, 2007 http://epw.senate.gov/public/index.cfm?FuseAction=Minority.Blogs&ContentRecord_id=922fe6c8-802a-23ad-49c5-25b74f6cdd7f

133 Bob Hebert, "Hot Enough Yet?" New York Times Op-Ed: August 3, 2006 http://groups.yahoo.com/group/laamn/message/9314

134 Alister Doyle, Polar Bears Drown, Islands Appear In Arctic Thaw, Reuters, September 15, 2006 http://www.wolfandwildlifestudies.com/news_article.php?id=2070

135 Senate Report, U.S. Senate Report Debunks Polar Bear Extinction Fears, January 30, 2008 http://epw.senate.gov/public/index.cfm?FuseAction=Minority.Blogs&ContentRecord_id=d6c6d346-802a-23ad-436f-40eb31233026

136 Future Retreat of Arctic Sea Ice Will Lower Polar Bear Populations and Limit Their Distribution, U.S. Geological Survey, September 7, 2006.

137 William Jasper "Polar Bears Thrive, Contrary to WWF Claims," The New American, January 13, 2009, http://www.thenewamerican.com/tech-mainmenu-30/environment/675

138 The ESA is the most effective federal tool to usurp local land use control and undermine private property rights. As landowners and businesses have known for decades, when you

want to stop a development project or just about any activity, find a species on that land to protect and things slow down or many times stop altogether. This is because Section 7 of the ESA requires that any project that involves the federal government in any way must meet the approval of the Fish and Wildlife Service before the project can move forward. This federal government involvement in a project can take the form of a federal grant, an environmental permit, a grazing allotment, a pesticide registration or land development permit. The law requires that the Fish and Wildlife Service intervene and determine if the project may affect an endangered or threatened species. So in the case of the polar bear listing, oil and gas exploration in Alaska, which accounts for 85% of the state's revenue and 25% of the nation's domestic oil production, is immediately called into question. Likewise, the state's shipping, highway construction, or fishing activities will be also be subject to federal scrutiny under Section 7. Part of this push was for listing the polar bear under the Endangered Species Act, was, of course, an attempt to exert some control over regulating greenhouse gases. The ESA is simply not equipped to regulate economy-wide greenhouse gases, nor does the Fish and Wildlife Service have the expertise to be a pollution control agency. The regulatory tools of the ESA function best when at-risk species are faced with local, tangible threats. Greenhouse gas emissions are not local. In essence, we can't scientifically establish a direct causal link from a CO2 molecule in Oklahoma or in Wyoming or in China to a direct effect on a polar bear in Alaska. We can't say which molecule is responsible. So how do you know who the culprit is and how do you regulate their activity under ESA?

139 Brian Montopoli, "Scott Pelley and Catherine Herrick On Global Warming Coverage, CBS News, March 23, 2006, http://www.cbsnews.com/8301-500486_162-1431768-500486.html.

140 Mark Whittington, "Doubting Global Warming Could Be Treason," Associated Content by Yahoo, July 16, 2007, http://www.associatedcontent.com/article/308598/doubting_global_warming_could_be_treason.html?cat=75.

141 David Roberts, "An Excerpt from a New Book by George Monbiot," Grist, September 19, 2006, http://www.grist.org/article/the-denial-industry.

142 Dr. Heidi Cullen, "Junk Controversy Not Junk Science...," Blog, Weather.com, December 21, 2006, http://www.weather.com/blog/weather/8_11392.html.

143 Editorial, "Doubting Inhofe," New York Times, October 12, 2006, http://www.nytimes.com/2006/10/12/opinion/12thu2.html.

144 Debra Saunders, "Inhofe the Apostate", San Francisco Chronicle, October 15, 2006 http://articles.sfgate.com/2006-10-15/opinion/17314756_1_global-warming-sen-james-inhofe-nasa-scientist-james-hansen

145 Jon Stewart On ClimateGate (VIDEO), Huffington Post, December 2, 2009 http://www.huffingtonpost.com/2009/12/02/jon-stewart-on-climategat_n_376672.html

146 Eric Pooley, Climategate Proves Sunlight Best Reply to Skeptics, Bloomberg, December 1, 2009 http://www.bloomberg.com/news/2009-12-01/climategate-proves-sunlight-best-reply-to-skeptics-eric-pooley.html

147 Garrison Keillor, Oped, The Love Weapon, New York Times, November 11, 2009 http://www.nytimes.com/2009/11/12/opinion/12iht-edkeillor.html

148 Scientists Hide Global Warming Data, The Daily Show, December 1, 2009 http://www.thedailyshow.com/videos/tag/Thomas%20Friedman

149 Senator Inhofe Responds to RFK Jrs "TRAITOR" Comments on FOX, Fox News with Neil Cavuto, July 10, 2007, http://www.youtube.com/watch?v=oheG6fTPwyU

150 Chris Mooney, The Republican War on Science, August 29, 2006 http://books.google.com/books?id=Jr5e7Tz5VNQC&pg=PT80&lpg=PT80&dq=%E2%80%9Cscience+abuser,%E2%80%9D+inhofe&source=bl&ots=wGrNgi8e_m&sig=WBxeT1nQkKUOKiiwvlHWC-ZibZQ&hl=en&ei=9RmuTrjSFYTb0QGotJ2UDw&sa=X&oi=book_result&ct=result&resnum=3&ved=0CCsQ6AEwAg#v=onepage&q=%E2%80%9Cscience%20

abuser%2C%E2%80%9D%20inhofe&f=false

151 Jim Myers, "Inhofe on 'enemies list,'" Tulsa World, January 13, 2010, http://www. tulsaworld.com/news/article.aspx?subjectid=16&articleid=20100113_16_A1_ OlhmeJ156867&archive=yes.

152 Matthew Sheffield, Senator Pounds Media for Global Warming Bias, Newsbusters, September 28, 2006. http://newsbusters.org/node/7972

153 CNN American Morning Transcript, January 31, 2007.

154 IPCC, 2007: Summary for Policymakers. In: Climate Change 2007: The Physical Science Basis. Contribution of Working Group I to the Fourth Assessment Report of the Intergovernmental Panel on Climate Change [Solomon, S., D. Qin, M. Manning, Z. Chen, M. Marquis, K.B. Averyt, M.Tignor and H.L. Miller (eds.)]. Cambridge University Press, Cambridge, United Kingdom
and New York, NY.

155 Steve McIntyre, IPCC Schedule: WG1 Report Available Only to Insiders Until May 2007, Climate Audit, January 24, 2007 http://climateaudit.org/2007/01/24/ipcc-4ar/

156 Lubos Motl, IPCC AR4, The Reference Frame, http://motls.blogspot.com/2007/01/ipcc-ar4. html January 25, 2007 http://motls.blogspot.com/2007/01/ipcc-ar4.html

157 Evangelical Leaders Exploited by Global Warming – Population Control Lobby, Real Catholic TV, September 29, 2006 http://www.realcatholictv.com/cia/04GlobalWarming/118.pdf

158 Transcript, "Could Global Warming Destroy Earth," Larry King Live, CNN, January 31, 2007, http://transcripts.cnn.com/TRANSCRIPTS/0701/31/lkl.01.html.

159 Senator Boxer Press Release, Boxer Announces Hearings on Clean Energy Jobs Bill to Start, October 27, 2009.

160 An Evangelical Declaration on Climate Change, Cornwall Alliance, 2009. http://www. cornwallalliance.org/articles/read/an-evangelical-declaration-on-global-warming/

161 Romans 1:25 NET, New English Translation, Biblical Studies Press, LLC, 1996-2007.

162 Bill Bright, "Promises, A Daily Guide to Supernatural Living," New Life Publications, 1998.

163 Transcript, CNN'S "AMERICAN MORNING," October 3, 2006.

164 Ben German, Gingrich calls climate ad with Pelosi the 'dumbest single thing I've done' recently," The Hill, November 9, 2011. http://thehill.com/blogs/e2-wire/e2-wire/192591-gingrich-calls-climate-ad-with-pelosi-the-dumbest-single-thing-ive-done

165 Transcript, CNN'S "AMERICAN MORNING," October 3, 2006.

166 Testimony of Dr. Robert Carter before the Senate Committee on Environment and Public Works, December 6, 2006 http://www.epw.senate.gov/109th/Carter_Testimony.pdf

167 Ibid.

168 Testimony of Dr. David Deming before the Senate Committee on Environment and Public Works, December 6, 2006 http://www.epw.senate.gov/hearing_statements.cfm?id=266543

169 Matthew Sheffield, Senate Debates Global Warming, CNN Anchor Snoozes, Newsbusters, December 6, 2006 http://newsbusters.org/people/television/miles-obrien?page=1

170 Kim Strassel, "Inhofe IMHO," Wall Street Journal, January 23, 2007, http://online.wsj.com/article/SB123186864579977861.html.

171 Emily Heil, "Drudge, Global Warming Shut Down Senate Site," The Hill, June 22, 2007, http://thehill.com/homenews/news/5668-drudge-global-warming-shut-down-senate-site.

172 Jonathan Hiskes, "Grading Senate Websites Reveals a Lack of Transarency on Climate and Energy," Grist, June 22, 2009, http://www.grist.org/article/2009-07-23-grading-senate-websites-climate-energy.

4: SKEPTICISM REIGNS

173 Charles Krauthammer, Column, "Limousine Liberal Hypocrisy," Time magazine, March 16, 2007, http://www.time.com/time/magazine/article/0,9171,1599714,00.html.

174 "With Five Private Jets, "Travolta Still Lectures on Global Warming," London Evening Standard, March 30, 2007, http://www.thisislondon.co.uk/showbiz/article-23390848-with-five-private-jets-travolta-still-lectures-on-global-warming.do.

175 Peter Schweizer, Op-ed, "Gore Isn't Quite as Green as He's Led the World to Believe," USA Today, December 7, 2006.

176 Al Gore, "An Inconvenient Truth," Documentary, 2006.

177 Senator Jim Inhofe, Blog, "Gore Refuses Pledge," Inhofe-EPW Website, March 21, 2007, http://epw.senate.gov/public/index.cfm?FuseAction=Minority.Pledge.

178 Transcript, "Countdown with Keith Olbermann,' for March 21, 2007" MSNBC, http://www.msnbc.msn.com/id/17737423/ns/msnbc_tv-countdown_with_keith_olbermann/t/countdown-keith-olbermann-march/.

179 Ibid

180 Ibid

181 Ibid

182 Kurt Orzeck, "A Global Divide? Reaction to Live Earth Is Decidedly Mixed," MTV, July 7, 2007, http://www.mtv.com/news/articles/1564253/reaction-live-earth-decidedly-mixed.jhtml.

183 Skeptical of Performers' Motives, Public Tunes Out Live Earth Event," Rasmussen, July 8, 2007, http://www.rasmussenreports.com/public_content/current_events/top_stories/skeptical_of_performers_motives_public_tunes_out_live_earth_event.

184 Ibid

185 Senator Inhofe Press Release, "Hollywood Celebrities Challenged to Take the 'Gore Pledge," Inhofe-EPW Website, April 20, 2007, http://epw.senate.gov/public/index.cfm?FuseAction=Minority.PressReleases&ContentRecord_id=10717b57-802a-23ad-400e-29c8948450c8.

186 "Countdown with Keith Olbermann," March 21, 2007" MSNBC, http://www.msnbc.msn.com/id/17737423/ns/msnbc_tv-countdown_with_keith_olbermann/t/countdown-keith-olbermann-march/.

187 Matt Crenson, "Gore Takes Heat in D.C. Return," Washington Post, March 24, 2007. http://www.washingtonpost.com/wp-dyn/content/article/2007/03/24/AR2007032401128.html

188 Al Gore, Op-Ed, "We Can't Wish Climate Change Away," New York Times, February 27, 2010. http://www.nytimes.com/2010/02/28/opinion/28gore.html?pagewanted=all

189 Alexander Bolton, Kerry: 'Dead wrong' to write obituary on climate change bill, The Hill, February 10, 2010 http://thehill.com/blogs/e2-wire/677-e2-wire/80539-kerry-those-writing-climate-legislation-obituary-are-dead-wrong

190 Brad Johnson, Inhofe's Grandchildren Build Igloo To Mock Killer Snow Storm: 'Al Gore's New Home', Think Progress, February 9, 2010 http://thinkprogress.org/politics/2010/02/09/81411/inhofe-family-gore-mockery/

191 Mark Asch, Jim Inhofe, America's Worst Senator, Also Has America's Worst Grandchildren, The Measure, February 10, 2010 http://www.thelmagazine.com/TheMeasure/archives/2010/02/10/jim-inhofe-americas-worst-senator-also-has-americas-worst-grandchildren

192 Tina Korbe, UA professor's ice storm story continues to snowball, The Arkansas Traveler, February 15, 2010 http://www.uatrav.com/2010/ua-professors-ice-storm-story-continues-to-snowball/

193 Dana Milbank, Global warming's snowball fight, Washington Post, February 14, 2010 http://www.washingtonpost.com/wp-dyn/content/article/2010/02/12/AR2010021203908.html

194 Ibid

195 Jeff Poor, Anatomy of a Botched Ambush: ABC Fails to Score Global Warming Point on Inhofe, Business and Media Institute, July 23, 2010 http://www.mrc.org/bmi/articles/2010/Anatomy_of_a_Botched_Ambush_ABC_Fails_to_Score_Global_Warming_Point_on_Inhofe.html

196 Americans Worried about Global Warming, Bush Not Doing Enough, Los Angeles Times/Bloomberg August 3,2006 http://environment.about.com/gi/o.htm?zi=1/XJ&zTi=1&sdn=environment&cdn=newsissues&tm=4&f=00&su=p284.12.336.ip_p504.1.336.ip_&tt=2&bt=1&bts=1&zu=http%3A//www.latimes.com/media/acrobat/2006-08/24711743.pdf

197 Claude Allegre, Op-Ed, The Snows of Mount Kilimanjaro, L'Express, September 21, 2006 http://epw.senate.gov/fact.cfm?party=rep&id=264835

198 Lawrence Solomon, The Real Deal, Canadian National Post, February 2, 2007 http://www.canada.com/nationalpost/story.html?id=156df7e6-d490-41c9-8b1f-106fef8763c6&k=0

199 Professor David Bellamy, Op-Ed, Global Warming? What a Load of Poppycock!, July 9, 2004 http://ff.org/centers/csspp/library/co2weekly/2005-04-28/load.htm

200 Dave Hoopman, "The Faithful Heretic," Wisconsin Energy Cooperative News, May 2007, http://www.wecnmagazine.com/2007issues/may/may07.html.

201 Letter, "Open Kyoto to debate," National Post, April 11, 2006, http://www.canada.com/nationalpost/financialpost/story.html?id=3711460e-bd5a-475d-a6be-4db87559d605.

202 President of Czech Republic Calls Man-Made Global Warming a 'Myth', Inhofe EPW Press Blog, February 12, 2007 http://epw.senate.gov/public/index.cfm?FuseAction=PressRoom.Blogs&ContentRecord_id=B6CD7713-802A-23AD-4AAF-A2D2ADDB287F

203 James Spann, Op-Ed, "The Weather Channel Mess," Blog, January 18, 2007, http://epw.senate.gov/public/index.cfm?FuseAction=PressRoom.Blogs&ContentRecord_id=3a9bc8a4-802a-23ad-4065-7dc37ec39adf.

204 Marc Morano, "Special Report: More Than 1,000 International Scientists Dissent over Man-Made Global Warming Claims," ClimateDepot.com, December 8, 2010, http://climatedepot.com/a/9035/SPECIAL-REPORT-More-Than-1000-International-Scientists-Dissent-Over-Man-made-Global-Warming-Claims--Challenge-UN-IPCC--Gore.

205 Juliet Eilperin, "Climate Is a Risky Issue for Democrats," Washington Post, November 6, 2007, http://www.washingtonpost.com/wp-dyn/content/article/2007/11/05/AR2007110502106.html.

206 Transcript, "Could Global Warming Destroy Earth," Larry King Live, CNN, January 31, 2007, http://transcripts.cnn.com/TRANSCRIPTS/0701/31/lkl.01.html.

207 Daniel Whitten, "Climate Bill Proving Vain Pursuit as Lobbies Roil U.S. Congress," Bloomberg, November 19, 2007, http://www.bloomberg.com/apps/news?pid=newsarchive&sid=aMRi__Kh4yKw&refer=uk.

208 Testimony from Anne E. Smith, Ph.D., Vice President, CRA International, Legislative Hearing on America's Climate Security Act of 2007, S.2191 of the Committee on Environment and Public Works, November 8, 2007, http://epw.senate.gov/public/index.cfm?FuseAction=Files.View&FileStore_id=80bc79be-c338-4a76-b438-205eb79da3d5.

209 Ibid.

210 Letter, American Federation of Labor and Congress of Industrial Organizations, November 5, 2007, http://www.hillheat.com/files/aflcio_gw_05.pdf.

211 Letter from Bruce Josten , Executive Vice President, Government Affairs, U.S. Chamber of Commerce to Honorable Joseph I. Lieberman and The Honorable John Warner , October 31, 2007, http://www.morpc.org/energy/green_region/Documents/Chamber.pdf.

212 Emily Pierce, "Global Warming Draws Heat from Democrats," Roll Call, June 3, 2008, http://www.rollcall.com/issues/53_145/-25560-1.html.

213 Environmental Protection Agency, EPA Analysis of the Lieberman-Warner Climate Security

Act of 2008, March 14, 2008 http://www.epa.gov/climatechange/downloads/s2191_EPA_Analysis.pdf

214 Energy Information Administration, Energy Market and Economic Impacts of S.2191, the Lieberman-Warner Climate Security Act of 2007, April 2008 http://www.eia.gov/oiaf/servicerpt/s2191/index.html

215 Science Applications International Corporation, Analysis of The Lieberman-Warner Climate Security Act (S.2191) Using the National Energy Modeling System (NEMS/ACCF/NAM) http://www.accf.org/pdf/NAM/fullstudy031208.pdf.

216 Energy Information Administration, Energy Market and Economic Impacts of S.2191, the Lieberman-Warner Climate Security Act of 2007, April 29, 2008 http://www.eia.gov/oiaf/servicerpt/s2191/index.html.

217 Ibid EIA.

218 Kate Sheppard, How the $6.7 trillion in Boxer's proposed amendment would be spent, Grist, May 21, 2008 http://www.grist.org/article/lieberman-warner-dolla-billz .

219 Congressional Budget Office, Approaches to Reducing Carbon Dioxide Emissions, Testimony before the Committee on the Budget, U.S. House of Representatives, November 1, 2007 http://www.cbo.gov/doc.cfm?index=8769.

220 CRA International, Economic Analysis of the Lieberman-Warner Climate Security Act of 2007 Using CRA's MRN-NEEM Model, April 8, 2008 http://www.crai.com/uploadedFiles/RELATING_MATERIALS/Publications/BC/Energy_and_Environment/files/CRA_NMA_S2191_April08_2008.pdf.

221 Senator Rockefeller Press Release, Rockefeller Frustrated by Failure of Senate to Act on Climate Change, June 6, 2008 http://rockefeller.senate.gov/press/record.cfm?id=298867

222 Stephen Moore, (Opinion) "Climate-Change Collapse," Wall Street Journal, June 6, 2008 http://online.wsj.com/article/SB121269184525849383.html?mod=googlenews_wsj.

5: THE MOMENT ARRIVES - AND THE MOVEMENT COLLAPSES

223 Transcript, "'The Rachel Maddow Show,' December 3, 2009," MSNBC, http://www.msnbc.msn.com/id/34275724/ns/msnbc_tv-rachel_maddow_show/t/rachel-maddow-show-thursday-december-rd/

224 David Weigel, "Inhofe on Maddow: 'I've Really Grown to Like that Gal,'" Washington Independent, December 8, 2009, http://washingtonindependent.com/70039/inhofe-on-maddow-ive-grown-to-like-that-gal.

225 Mark Lynas, "Yvo de Boer's resignation compounds sense of gathering climate crisis," Guardian, February, 18, 2010, http://www.guardian.co.uk/environment/2010/feb/18/yvo-de-boer-resignation-un-climate-change-body .

226 Juliet Eilperin, "An Inconvenient Expert," Outside Magazine, September 28, 2007, http://www.outsideonline.com/outdoor-adventure/An-Inconvenient-Expert.html

227 Erika Lovely, "Scientists urge caution on global warming," Politico, November 25, 2008, http://dyn.politico.com/printstory.cfm?uuid=D0C4924D-18FE-70B2-A808D77A9C1FFFD3 .

228 Lorne Gunter, (Opinion), "Skeptics, unite!," National Post, October 20, 2008, http://www.nationalpost.com/opinion/story.html?id=bee2f7de-cb35-4f3d-9652-48c97f8ae623

229 Andrew Revkin, "Counting Skeptics" Comment on Grist Blog, , March 6, 2008, http://www.grist.org/article/four-hundred-skeptics-try-19 .

230 Press Conference by the President, June 23, 2009, http://www.whitehouse.gov/the-press-office/press-conference-president-6-23-09.

231 Senator Boxer, Opening Statement, Senate Committee on Environment and Public Works Committee Hearing, Full Committee Hearing Entitled, "Moving America Toward a Clean

Energy Economy and Reducing Global Warming Pollution: Legislative Tools." July 7, 2009, http://epw.senate.gov/public/index.cfm?FuseAction=Hearings.Statement&Statement_ID=5b78ff44-9cc5-4914-ba8c-003d34d62d00

232 John M. Broder, "Seeking to Save the Planet, With a Thesaurus," New York Times, May 1, 2009, http://www.nytimes.com/2009/05/02/us/politics/02enviro.html .

233 Peter Nicholas, "Buzzwords: Rephrasing Obama's Lexicon," Los Angeles Times, May 11, 2009, http://articles.latimes.com/2009/may/11/nation/na-obama-language11 .

234 Allison Winter, Boxer, Kerry Say Comprehensive Bill is on Track for Wednesday, Environment and Energy Daily, September 25, 2009 http://www.eenews.net/EEDaily/2009/09/25/2/

235 Glenn Thrush, "Dingell: Cap and trade a 'great big' tax" Politico, April 27, 2009, http://www.politico.com/news/stories/0409/21730.html.

236 Henry Payne, "Sen. Debbie Stabenow, Energy Leader," The Detroit News, August 10, 2009 http://community.detnews.com/apps/blogs/henrypayneblog/index.php?blogid=2041

237 Wayne Trotter, "Inhofe Says 'Cap And Trade' Dead," Countywide & Sun, June 26, 2009, http://countywidenews.com/inhofe-says-cap-and-trade-dead-p1352.htm.

238 Senator Inhofe Press Release, "Jackson Confirms EPA Chart Showing No Effect on Climate without India, China," July 7, 2009, http://epw.senate.gov/public/index.cfm?FuseAction=Minority.PressReleases&ContentRecord_id=564ed42f-802a-23ad-4570-3399477b1393&Region_id=&Issue_id= .

239 Richard S. Dunham, Obama defends climate bill's cap-and-trade plan, The San Francisco Chronicle, June 25, 2009 http://www.sfgate.com/cgi-bin/article.cgi?f=/c/a/2009/06/24/MNF518DBGI.DTL

240 U.S. Senate Environment and Public Works Committee Hearing, Full Committee and Subcommittee on Green Jobs and the New Economy joint hearing entitled, "Clean Energy Jobs, Climate-Related Policies and Economic Growth - State and Local Views." Senator Inhofe Press Release, "Colorado Gov Refuses to Endorse Waxman-Markey," July 21, 2009, http://epw.senate.gov/public/index.cfm?FuseAction=PressRoom.PressReleases&ContentRecord_id=9E652ED3-802A-23AD-4238-2170377B26FD .

241 Harry Alford, Testimony Before the Senate Committee on Environment and Public Works, Ensuring and Enhancing U.S. Competitiveness While Moving Toward a Clean Energy Economy July 16, 2009 http://epw.senate.gov/public/index.cfm?FuseAction=Files.View&FileStore_id=3f34f593-8c87-46f2-bba7-f622481ecfb2.

242 Senator Inhofe Press Release, "Boxer Says Goal is to Soften the Blow," July 16, 2009, http://epw.senate.gov/public/index.cfm?FuseAction=Minority.PressReleases&ContentRecordType_id=ae7a6475-a01f-4da5-aa94-0a98973de620&Region_id=&Issue_id=&MonthDisplay=7&YearDisplay=2009.

243 A closer look at the bill revealed Title IV, Section B, Part 2, is called "Climate Change Worker Adjustment Assistance." Just beneath that is Sec. 425, called "Petitions, Eligibility Requirements, And Determinations." This provision allowed workers to file for a "certification of eligibility" as a group with the Department of Labor. These workers could apply for "adjustment assistance," subsequent to a hearing to determine if they are eligible. Then there was this suspicious part in Sec. 426 called "Program Benefits" that allowed for a "climate change adjustment allowance" for an "adversely affected worker." The obvious question here was: "adversely affected" by what? By the bill, of course. This provision authorizes payment to these workers for a week of unemployment that "shall equal 70% of the average weekly wage of that worker for a period of not longer than 156 weeks," or 3 years. Reading on there were also job training benefits, including "individual career counseling" and "prevocational services," defined as "development of learning skills, communications skills, interviewing skills, punctuality, personal maintenance skills, and professional conduct to prepare individuals for employment or training." But that's not all: workers may be eligible in certain

circumstances for a one-time job search allowance up to $1,500, and for relocation assistance up to $1,500. So in specifically adding provisions for the unemployed the authors themselves were implicitly acknowledging that Waxman-Markey would destroy jobs.

244 Congressional Research Service, U.S. Fossil Fuel Resources: Terminology, Reporting, and Summary, October 28, 2009 http://epw.senate.gov/public/index.cfm?FuseAction=Files. View&FileStore_id=f7bd7b77-ba50-48c2-a635-220d7cf8c519

245 Darren Samuelsohn, "Republicans on Track to Boycott Senate EPW Panel's Markup," Environment and Energy Daily, October 30, 2009, http://www.eenews.net/eenewspm/print/2009/10/30/1.

246 America's News Room, Fox News, November 5, 2009.

247 Ed Morrissey, Inhofe to Boxer: "We won, you lost – get a life!" Hot Air, November 19, 2009 http://hotair.com/archives/2009/11/19/inhofe-to-boxer-we-won-you-lost-get-a-life/

6: CLIMATEGATE = VINDICATION

248 Dana Milbank, "Washington Sketch: A hostile climate for Sen. Inhofe the warming skeptic," Washington Post, October 28, 2009, http://www.washingtonpost.com/wp-dyn/content/article/2009/10/27/AR2009102702845.html .

249 Louise Gray, "Analysis: Copenhagen is going to be little more than a photo opportunity," Telegraph, http://www.telegraph.co.uk/earth/6576232/Analysis-Copenhagen-is-going-to-be-little-more-than-a-photo-opportunity-for-world-leaders.html .

250 "Inhofe Calls 2009 The Year of the Skeptic In Senate Floor Speech," U.S. Senate Committee on Environment & Public Works, November 18, 2009, http://epw.senate.gov/public/index.cfm?FuseAction=Minority.PressReleases&ContentRecord_id=0a725d63-802a-23ad-44ea-01cf57e06fb7.

251 Senate Report: 'Consensus' Exposed: The CRU Controversy, February 2010 http://epw.senate.gov/public/index.cfm?FuseAction=Files.View&FileStore_id=7db3fbd8-f1b4-4fdf-bd15-12b7df1a0b63

252 Naomi Oreskes, "Beyond the Ivory Tower: The Scientific Consensus on Climate Change," Science 306, no. 5702 (December 3, 2004): 1686, http://www.sciencemag.org/content/306/5702/1686.full .

253 Testimony of Dr. Jane Lubchenco, Undersecretary of Commerce for Oceans and Atmosphere and NOAA Administrator National Oceanic and Atmospheric Administration U.S. Department of Commerce, Hearing on "The Administration's View on the State of Climate Science, December 2, 2009, http://www.legislative.noaa.gov/Testimony/Lubchenco120209.pdf .

254 Report, Climate Change 2007: Synthesis Report, Summary for Policymakers, http://www.ipcc.ch/pdf/assessment-report/ar4/syr/ar4_syr_spm.pdf .

255 George Monbiot, "Pretending the climate email leak isn't a big crisis won't make it go away," Guardian UK, Wednesday, November 25, 2009, http://www.guardian.co.uk/environment/georgemonbiot/2009/nov/25/monbiot-climate-leak-crisis-response .

256 James Delingpole, Climategate: the final nail in the coffin of 'Anthropogenic Global Warming'?, Telegraph, November 20, 2009 http://blogs.telegraph.co.uk/news/jamesdelingpole/100017393/climategate-the-final-nail-in-the-coffin-of-anthropogenic-global-warming/

257 Clive Crook, A heated debate, Comment Posted at the Economist, December 1, 2009 https://www.economist.com/user/3476265/comments

258 Jon Stewart, "Scientists Hide Global Warming Data," The Daily Show, Comedy Central, December 1, 2009

259 Senator Inhofe speech, Bringing Integrity Back to the IPCC Process, November 15, 2005 http://epw.senate.gov/public/index.cfm?FuseAction=Minority.Speeches&ContentRecord_

id=21cc88ec-cca6-4a61-8c2e-78fa8de4850d&Region_id=&Issue_id=

260 Evidence is now 'unequivocal' that humans are causing global warming – UN report, UN News Centre, February 2, 2007 http://www.un.org/apps/news/story.asp?Cr1=change&Cr=climate&NewsID=21429

261 Robert Mendick, Climategate: University of East Anglia U-turn in climate change row, Telegraph, November 28, 2009, http://www.telegraph.co.uk/earth/copenhagen-climate-change-confe/6678469/Climategate-University-of-East-Anglia-U-turn-in-climate-change-row.html .

262 Review & Outlook, "A Climate Absolution? The alarmists still won't separate science from politics." Wall Street Journal. July 16, 2010. http://online.wsj.com/article/SB1000142405 2748703394204575367483847033948.html

263 University of East Anglia emails: the most contentious quotes, UK Telegraph, November 23, 2009 http://www.telegraph.co.uk/earth/environment/globalwarming/6636563/University-of-East-Anglia-emails-the-most-contentious-quotes.html .

264 See, United States Senate Environment and Public Works Minority Staff Report, "'Consensus' Exposed: The CRU Controversy," February 2010, http://epw.senate.gov/public/index.cfm?FuseAction=Files.View&FileStore_id=7db3fbd8-f1b4-4fdf-bd15-12b7df1a0b63

265 "20th Century Climate Not So Hot," Press Release, Harvard-Smithsonian Center for Astrophysics, March 31, 2003, http://www.cfa.harvard.edu/news/archive/pr0310.html

266 United States Senate Environment and Public Works Minority Staff Report, February 2010.

267 Ibid

268 Ibid

269 Ibid

270 Ibid

271 Quirin Schiermeier, Storm clouds gather over leaked climate e-mails, Published online November 24, 2009 at http://www.nature.com/news/2009/091124/full/462397a.html

272 Dr. John Holdren, hearing, House Select Committee on Energy Independence and Global Warming, December 2, 2009, http://www.noaanews.noaa.gov/stories2009/pdfs/cqtranscript_dec2houseclimatehearing.pdf .

273 Gerald Traufetter, Confidence Melting Away. Can Climate Forecasts Still Be Trusted? Speigel Online January 27, 2010. http://www.spiegel.de/international/world/0,1518,674087-2,00.html

274 See National Research Council, "Surface Temperature Reconstructions for the Last 2,000 Years," Committee on Surface Temperature for the Last 2,000 Years, 22 June 2006.

275 United States Senate Minority Staff Report, February 2010

276 Ibid

277 Ibid

278 David Rose, "SPECIAL INVESTIGATION: Climate change emails row deepens as Russians admit they DID come from their Siberian server," Mail Online, December 13, 2009, http://www.dailymail.co.uk/news/article-1235395/SPECIAL-INVESTIGATION-Climate-change-emails-row-deepens--Russians-admit-DID-send-them.html

279 Ibid.

280 Massachusetts v. EPA, 549 U.S. 497, 127 S. Ct. 1438 (2007).

281 Editorial, Reckless 'Endangerment' Wall Street Journal, April 24, 2009 http://online.wsj.com/article/SB124052921804450391.html

282 Darren Samuelsohn, Prospect of EPA regulations a 'glorious mess' – Dingell, Greenwire, April 8, 2008 http://www.eenews.net/public/Greenwire/2008/04/08/1

283 Lisa Lerer, Republicans Push On Climategate, Politico, December 6, 2009 http://www.politico.com/news/stories/1209/30172_Page2.html

284 Chris Casteel, U.S. Sen. Jim Inhofe challenges EPA's climate change policies, Daily Oklahoman, February 24, 2010 http://newsok.com/sen.-jim-inhofe-challenges-epas-climate-

change-policies/article/3441792?custom_click=lead_story_title

285 EPA Endangerment Finding, 74 Fed. Reg. 66511, December 15, 2009

286 Technical Support Document for Endangerment Finding and Cause or Contribute Findings for Greenhouse Gases under Section 202(a) of the Clean Air Act, December 7, 2009, 47

287 Alan Carlin, Comments on Draft Technical Support Document for Endangerment Analysis for Greenhouse Gas Emissions under the Clean Air Act, March 16, 2009.

288 Dr. Alan Carlin, Proposed NCEE Comments on Draft Technical Support Document for Endangerment Analysis for Greenhouse Gas Emissions under the Clean Air Act, March 9. 2009, http://cei.org/cei_files/fm/active/0/DOC062509-004.pdf)

289 The Environmental Protection Agency, Endangerment and Cause or Contribute Findings for Greenhouse Gases under Section 202(a) of the Clean Air Act http://epa.gov/climatechange/endangerment.html

290 Gerald Traufetter, Can Climate Forecasts Still Be Trusted? ABC News, January 28, 2010 http://epw.senate.gov/public/index.cfm?FuseAction=Minority.Blogs&ContentRecord_id=7633F291-802A-23AD-42E8-B295A147D2B1

291 Off-base camp, The Economist, January 21, 2010 http://www.economist.com/node/15328534?story_id=15328534

292 Bryan Walsh, Himalayan Melting: How a Climate Panel Got It Wrong, Time Magazine, January 21, 2010 http://www.time.com/time/health/article/0,8599,1955405,00.html#ixzz1aWnXy6cw

293 Editorial, Uncertain Science, Newsweek, May 27, 2010 http://www.thedailybeast.com/newsweek/2010/05/28/uncertain-science.html

294 Editorial, Time for a change in climate research, Financial Times, August 31, 2010 http://www.ft.com/intl/cms/s/0/362ddb00-b534-11df-9af8-00144feabdc0.html#axzz1cKkXJQcJ

295 Richard Alleyne, UN report on glaciers melting is based on 'speculation', Telegraph, January 17, 2010 http://www.telegraph.co.uk/earth/earthnews/7011713/UN-report-on-glaciers-melting-is-based-on-speculation.html

296 Ibid

297 Bryan Walsh, Himalayan Melting: How a Climate Panel Got It Wrong, Time Magazine, January 21, 2010 http://www.time.com/time/health/article/0,8599,1955405,00.html#ixzz1aWnXy6cw

298 The Telegraph, "Climate change leader 'knew about false glacier claims before Copenhagen.'" January 31, 2010. http://www.telegraph.co.uk/earth/environment/climatechange/7107873/Climate-change-leader-knew-about-false-glacier-claims-before-Copenhagen.html

299 David Rose, Glacier scientist: I knew data hadn't been verified, Daily Mail, January 24, 2010, http://www.dailymail.co.uk/news/article-1245636/Glacier-scientists-says-knew-data-verified.html

300 IPCC statement on the melting of Himalayan glaciers, January 20, 2010 http://www.ipcc.ch/pdf/presentations/himalaya-statement-20january2010.pdf

301 Richard Gray, UN climate change panel based claims on student dissertation and magazine article, Telegraph, Jan 30, 2010, http://www.telegraph.co.uk/earth/environment/climatechange/7111525/UN-climate-change-panel-based-claims-on-student-dissertation-and-magazine-article.html

302 Office of the Inspector General, Environmental Protection Agency, Procedural Review of EPA's Greenhouse Gases Endangerment Finding Data Quality Processes, September 26, 2011, http://www.epa.gov/oig/reports/2011/20110926-11-P-0702.pdf

303 Dina Cappiello, Report: EPA Cut Corners on Climate Finding, Associated Press, September 28, 2011 http://www.guardian.co.uk/world/feedarticle/9869682

7: COPENHAGEN

304 William La Jeunesse, "Sen. Inhofe (R) In Lion's Den in Copenhagen," Fox News, December 17, 2009, http://liveshots.blogs.foxnews.com/2009/12/17/sen-inhofe-r-in-lions-den-in-copenhagen/

305 Christian Schwägerl, "Copenhagen Reveals a Vicious Circle of Mistrust" Der Spiegel Editorial, December 21, 2009, http://www.spiegel.de/international/world/0,1518,668408,00.html

306 President Obama, Remarks by the President at United Nations Secretary General Ban Ki-Moon's Climate Change Summit, September 22, 2009, http://m.whitehouse.gov/the_press_office/Remarks-by-the-President-at-UN-Secretary-General-Ban-Ki-moons-Climate-Change-Summit/

307 "Copenhagen's Collapse: The climate change sequel is a bust," Wall Street Journal Editorial, November, 17, 2009, http://online.wsj.com/article/SB10001424052748704431804574540002267533772.html

308 Anna Fifield, "US senator calls global warming a 'hoax,'" Financial Times, December 4, 2009, http://www.ft.com/intl/cms/s/0/b68c23ea-e0fc-11de-af7a-00144feab49a.html#axzz1cIAlHRUB

309 "Transcript: Sen. Inhofe, Rep. Markey on 'FNS,'" Fox News Sunday with Chris Wallace, December 14, 2009, http://www.foxnews.com/on-air/fox-news-sunday/2009/12/14/transcript-sen-inhofe-rep-markey-fns

310 "Senator Kerry Smacks Down Senator Inhofe in Copenhagen," YouTube Video, December 16, 2009, http://www.youtube.com/watch?v=JLwaLTT5ZxI

311 Inhofe Speech, Inhofe in Copenhagen: "It Has Failed ... It's Déjà Vu All Over Again.", December 17, 2009 http://epw.senate.gov/public/index.cfm?FuseAction=Minority.PressReleases&ContentRecord_id=0f7106cd-802a-23ad-497a-880aa69d1a7f&Region_id=&Issue_id=

312 Transcript, The Situation Room, CNN, December 18, 2009 http://transcripts.cnn.com/TRANSCRIPTS/0912/18/sitroom.01.html

313 Sharyl Attkisson, "Copenhagen Summit Turned Junket?" CBS, January 20, 2010, http://www.cbsnews.com/stories/2010/01/11/cbsnews_investigates/main6084364.shtml?tag=contentBody;featuredPost-PE

8: THE ATTEMPTS TO RAISE CAP-AND-TRADE FROM THE GRAVE

314 Lou Dobbs, "Inhofe Deserves a Great Deal of Applause and Respect and Has Been Vindicated on Global Warming," Lou Dobbs Radio Show, YouTube, February 24, 2010 http://epw.senate.gov/public/index.cfm?FuseAction=Minority.Blogs&ContentRecord_id=01bd94bc-802a-23ad-4e59-738d7c3ba83e&Issue_id=

315 David A. Fahrenthold, "Kerry, Graham, Lieberman announce a 'dual track' on the climate bill," Washington Post, November 4, 2009 http://voices.washingtonpost.com/capitol-briefing/2009/11/kerry_graham_lieberman_announc.html

316 Steve Benen, "Kerry, Graham, Lieberman Release Climate Bill 'Framework,'" Washington Monthly, December 10, 2009 http://www.washingtonmonthly.com/archives/individual/2009_12/021401.php

317 John Kerry and Lindsey Graham, Op-Ed, "Yes We Can (Pass Climate Change Legislation)," New York Times Op Ed, October 10, 2009 http://www.nytimes.com/2009/10/11/opinion/11kerrygraham.html?pagewanted=all

318 Benen, Washington Monthly.

319 Darren Samuelsohn, "Dems look for optimism on energy vote after vote to preserve EPA cli-

mate regs," Environment and Energy Daily, June 11, 2010, http://www.eenews.net/EEDaily/print/2010/06/11/1

320 Interview with Senator Barack Obama, San Francisco Chronicle, January 2008.

321 Jim Tankersley, Senators consider gasoline tax as part of climate bill, Los Angeles Times, April 14, 2010 http://articles.latimes.com/2010/apr/14/nation/la-na-gas-tax14-2010apr14

322 WH Opposes High Gas Taxes Floated by S.C. GOP Sen. Graham in Emerging Senate Energy Bill, Fox News, April 15, 2010 http://politics.blogs.foxnews.com/2010/04/15/wh-opposes-higher-gas-taxes-floated-sc-gop-sen-graham-emerging-senate-energy-bill

323 Jeanne Cummings, Inhofe: Climate's no better for bill, Politico, April 19, 2010 http://www.politico.com/news/stories/0410/35967.html

324 Linda Feldmann,"Obama wants 'price' on carbon emissions. Republicans see 'tax." Christian Science Monitor, June 29, 2010. http://www.csmonitor.com/USA/Politics/2010/0629/Obama-wants-price-on-carbon-emissions.-Republicans-see-tax.

325 Brief for Respondents at 49, Coalition for Responsible Regulation, Inc., et al. v. United States Environmental Protection Agency, no.10-1073 (D.C. Cir. Sept. 16, 2011)

326 Alexander Bolton, "Climate change showdown," The Hill, June 9, 2010, http://thehill.com/homenews/senate/102377-climate-change-showdown

327 Statement of Rep. Henry A. Waxman, Ranking Member, Committee on Energy and Commerce, "H.R. ___, the Energy Tax Prevention Act of 2011," Subcommittee on Energy and Power, February 9, 2011. http://democrats.energycommerce.house.gov/sites/default/files/image_uploads/Waxman_OpeningStatement.pdf). He went on to say that our bill would "repeal the only authority the administration has to protect our health and the environment without providing any alternative."

328 Ibid.

329 Amy Harder, "Reid Might Be Forced to Deal with His EPA Problem," National Journal, March 28, 2010, http://www.nationaljournal.com/member/daily/reid-might-be-forced-to-deal-with-his-epa-problem-20110328?print=true

330 Letter to President Obama, February 28, 2011, http://brown.senate.gov/newsroom/press_releases/release/?id=15d01a11-40f3-4221-9a7b-c7bd498ec372

331 Robin Bravender, "Democrats move on from cap-and-trade," Politico, November 17, 2010, http://www.politico.com/news/stories/1110/45241.html

332 President Obama, "Remarks by the President in State of Union Address," January 25, 2011, http://www.whitehouse.gov/the-press-office/2011/01/25/remarks-president-state-union-address

333 Obama Administration, "Chapter Six: Transitioning to a Clean Energy Future," Economic Report of the President, February 2011, http://www.whitehouse.gov/sites/default/files/microsites/2011_erp_chapter6.pdf

334 Kimberley A. Strassel, "Cap and Trade Returns from the Grave," Wall Street Journal, January 28, 2011, http://online.wsj.com/article/SB10001424052748703893104576108501552298070.html#printMode

335 President Obama, "Remarks of President Barack Obama, Address to Joint Session of Congress," February 24, 2009, http://www.whitehouse.gov/the_press_office/Remarks-of-President-Barack-Obama-Address-to-Joint-Session-of-Congress/

336 United States Senate Report, "The Real Story Behind China's Energy Policy," December 8, 2010 http://epw.senate.gov/public/index.cfm?FuseAction=Files.View&FileStore_id=f29ee5f7-c9f5-46ca-9500-0f10b27f41ed

337 President Obama, "Full Text of President Barack Obama's speech at Penn State," February 3, 2011 http://live.psu.edu/story/51187

338 President Obama, "Remarks by the President on the Economy," Solyndra, Inc., May 26, 2010 http://www.whitehouse.gov/the-press-office/remarks-president-economy-0

339 "Vice President Biden Announces Finalized $535 Million Loan Guarantee for Solyndra," Department of Energy Press Release, September 4, 2009, http://energy.gov/articles/vice-president-biden-announces-finalized-535-million-loan-guarantee-solyndra

340 Congressional Research Service, U.S. Fossil Fuel Resources: Terminology, Reporting, and Summary (R40872), March 25, 2011, (update requested by Senator James M. Inhofe). http://www.crs.gov/pages/Reports.aspx?PRODCODE=R40872&Source=search#Content

EPILOGUE: GETTING OUR COUNTRY BACK ON TRACK

341 Jim Myers, "Inhofe believes swimming in Grand Lake cause of his illness," Tulsa World, July 1, 2011, http://www.tulsaworld.com/news/article.aspx?subjectid=335&articleid=20110701_335_0_hrimgs611815

342 "Sierra Club to Inhofe: Get well (and green) soon," Politico, July 7, 2011, http://www.politico.com/blogs/glennthrush/0711/Sierra_Club_to_Inhofe_Get_well_and_green_soon.html

343 Stephen Stromberg, A funny thing happened at the climate denier conference, Washington Post, June 20, 2011 http://www.washingtonpost.com/blogs/post-partisan/post/a-funny-thing-happened-at-the-climate-denier-conference/2011/06/30/AGLOmEsH_blog.html

344 Senator Inhofe, Inhofe: "Nothing is going to happen in Cancun at UN Climate party and everyone knows it," Inhofe EPW Press Release, December 3, 2010, http://epw.senate.gov/public/index.cfm?FuseAction=Minority.Blogs&ContentRecord_id=ac7bbe30-802a-23ad-4d13-f64b2af51cfb

345 Jean Chemnick, New Batch of Hacked Emails Released Ahead of Durban Talks, Greenwire, November 22, 2011 http://www.eenews.net/Greenwire/2011/11/22/archive/4?terms=New+batch+of+hacked+emails+released+ahead+of+Durban+talks+)

346 James Hansen, Op-Ed, "Cap and Fade," New York Times, December 6, 2009, http://www.nytimes.com/2009/12/07/opinion/07hansen.html

347 Tracy McVeigh, "Backlash over Richard Curtis's 10:10 climate film," Guardian UK, October 2, 2010, http://www.guardian.co.uk/environment/2010/oct/02/1010-richard-curtis-climate-change.

348 Damian Carrington, "There will be blood – watch exclusive of 10:10 campaign's 'No Pressure' film," Guardian UK Blog, September 30, 2010, http://www.guardian.co.uk/environment/blog/2010/sep/30/10-10-no-pressure-film.

349 See, http://berkeleyearth.org/study.php for information on the study. Berkeley Earth Surface Temperature, 2011.

350 Richard A. Muller , "The Case Against Global-Warming Skepticism " Wall Street Journal Op Ed, October 21, 2011, http://online.wsj.com/article/SB10001424052970204422404576594872796327348.html#printMode

351 Seth Borenstein, "Skeptic finds he now agrees global warming is real," AP, Oct 31, 2011., http://news.yahoo.com/skeptic-finds-now-agrees-global-warming-real-142616605.html.

352 David Rose, "Scientist who said climate change skeptics had been proved wrong accused of hiding truth by colleague," Mail Online, October 30, 2011, http://www.dailymail.co.uk/sciencetech/article-2055191/Scientists-said-climate-change-sceptics-proved-wrong-accused-hiding-truth-colleague.html .

353 Ibid.

354 Robert Gammon, "Friday Must Read: Boxer Says Cap-and-Trade is Dead; Dellums to Finish Term," East Bay Express, November 19, 2010, http://www.eastbayexpress.com/92510/archives/2010/11/19/friday-must-read-boxer-says-cap-and-trade-is-dead-dellums-to-finish-term

355 Ben German, "Gingrich calls climate ad with Pelosi the 'dumbest single thing I've done'

recently," The Hill, November 9, 2011. http://thehill.com/blogs/e2-wire/e2-wire/192591-gingrich-calls-climate-ad-with-pelosi-the-dumbest-single-thing-ive-done

356 EPA Inspector General, "Procedural Review of EPA's Greenhouse Gases Endangerment Finding Data Quality Processes Report No. 11-P-0702, September 26, 2011, http://www.epa.gov/oig/reports/2011/20110926-11-P-0702.pdf.

357 Bill Bright, Promises, A Daily Guide to Supernatural Living, New Life Publications, 1998.

AFTERWORD: WHAT GLOBAL WARMING AND EARMARKS HAVE IN COMMON

358 John Stanton, "Inhofe is Happy to Stand Apart," Roll Call, November 22, 2010, http://www.rollcall.com/issues/56_50/-200810-1.html.

359 Paul Weyrich, "Senator Inhofe: Transportation, work and achievement," Renew America, September 12, 2006, http://www.renewamerica.com/columns/weyrich/060912

360 Statement of Rep. Henry A. Waxman, Ranking Member, Committee on Energy and Commerce, "H.R. ___, the Energy Tax Prevention Act of 2011," Subcommittee on Energy and Power, February 9, 2011. http://democrats.energycommerce.house.gov/sites/default/files/image_uploads/Waxman_OpeningStatement.pdf

361 Ibid.

362 Massachusetts v. EPA, 549 U.S. 497, 127 S. Ct. 1438 (2007).

363 Opening Statement of Lisa P. Jackson, Administrator United States Environmental Protection Agency, Hearing on a Draft Bill to Eliminate Portions of the Clean Air Act, Subcommittee on Energy and Power, Committee on Energy and Commerce, U.S. House of Representatives, Feb. 9, 2011. http://republicans.energycommerce.house.gov/Media/file/Hearings/Energy/020911_Energy_Tax_Prevention_Act/Jackson.pdf

364 Public Law 110-140

365 Maeve P. Carey and Betsy Palmer, "Presidential Appointments, the Senate's Confirmation Process, and Proposals for Change, 112th Congress," Congressional Research Service, July 8, 2011.

366 Ed O'Keefe and Eric Yoder, "Boehner's comments revive debate on how to tally federal workers." Washington Post, Feb. 17, 2011. http://www.washingtonpost.com/wp-dyn/content/article/2011/02/16/AR2011021606846.html

367 Waste 102: Harvard's Robotic Bee Project Tops List of Government's Most Reckless Spending," Fox News,March 15, 2010, http://www.foxnews.com/on-air/hannity/2010/03/15/waste-102-harvards-robotic-bee-project-tops-list-governments-most-reckless-spending

368 U.S. Senator James M. Inhofe, Congressional Record, March 15, 2010, Pp. S1494-1496.

369 Eric Lichtblau, "Leaders in House Block Earmarks to Corporations," New York Times, March 10, 2010.http://www.nytimes.com/2010/03/11/us/politics/11earmark.html

370 Scott Wong, "Senate Dems give in on earmark ban," Politico, February 1, 2011 http://www.politico.com/news/stories/0211/48623.html

371 Blake Morrison, Tom Vanden Brook, and Peter Eisler, "Congress Intervened to Protect Troops from IEDs," USA Today, Oct. 2, 2007. http://www.usatoday.com/news/military/2007-09-03-congressmrap_N.htm

372 Rules of the House of Representatives, 112th Congress, Jan. 5, 2011. http://clerk.house.gov/legislative/house-rules.pdf

373 Jonathan Rauch, Earmarks are a Model Not a Menace, National Journal, March 14, 2009,

374 U.S. Senate Armed Services Committee, "National Defense Authorization Act for Fiscal Year 2011," Report to Accompany S. 3454, U.S. Senate Report 111-201, 111th Congress, 2nd Session. June 4, 2010.

375 James M. Inhofe, Op-ed, Washington Times, December 3, 2010.

376 James M. Inhofe and Ron Paul, Earmark Ban a Huge Victory for Obama, Human Events, 3/11/2011, http://www.humanevents.com/article.php?id=42234

377 Senate Resolution 23 (112th Congress), "A resolution to prohibit unauthorized earmarks." Introduced Jan. 25, 2011.

378 U.S. Senator John McCain, Congressional Record, pg. S8233, Nov. 29, 2010.

379 U.S. Senator Tom Coburn, Congressional Record, pg. S8232, Nov. 29, 2010.

380 Merriam Webster's Collegiate Dictionary, 10th Edition, 1993.

APPENDIX A: WHAT'S IN IT FOR THE UNITED NATIONS?

381 United Nations, UN At a Glance, accessed 10/24/2011 from www.un.org/en/aboutun/index.shtml

382 UN Charter, Article 1, Section 1

383 UN Charter, Article 1, Section 2

384 UN Charter, Article 1, Section 3

385 UN Charter, Article 1, Section 4

386 Unless otherwise specified, when I refer to the UN I am referring to the General Assembly. The Security Council, which can make binding resolutions, is structured differently from the General Assembly.

387 UN Charter, Article 18, Paragraph 1.

388 United Nations, Overview of the UN Global Compact, www.unglobalcompact.org/AboutTheGC

389 World Commission on Environment and Development, August 4, 1987, Our Common Future, U.N. Doc. A/42/427

390 Al Gore, Earth in the Balance, 366 (1992)

391 Al Gore, Earth in the Balance, 368 (1992)

392 Richard A Matthew & Anne Hammill, Sustainable Development and Climate Change, International Affairs, 85, 1117 (2009).

393 Al Gore, Earth in the Balance, 3 (1992)

394 World Commission on Environment and Development, August 4, 1987, Our Common Future, U.N. Doc. A/42/427, 55

395 Ibid, 54-55

396 Maurice Strong, Where on Earth Are We Going? (2000), Page 121

397 G77.org.

398 Maurice Strong, Where on Earth Are We Going? (2000), Page 124.

399 Branislav Gosovic, the Quest for World Environmental Cooperation, Page 6

400 Ibid

401 G.A. Res. 3201, U.N. Doc. A/9559 (May 1, 1974), page 3.

402 UNEP/UNCTAD Symposium on "Patterns of Resource Use, Environment, and Development Strategies", The Cocoyoc Declaration, U.N. Doc. A/C.2/292 (November 1, 1974), page 2.

403 Al Gore, Earth in the Balance, 274 (1992)

404 World Summit for Social Development, Report of the World Summit for Social Development, U.N.Doc.A/CONF.166/9 (April 19, 1995)

405 UN Press Release on the United Nations Conference on Trade and Development, U.N. Doc. HAB/IST/25 (June 15, 1996)

406 G.A. Res. S-27/2, U.N. Doc. A/RES/S-27/2 (October 11, 2002)

407 Carolyn Dimitri, Anne Effland, and Neilson Conklin, The 20th Century Transformation of U.S. Agriculture and Farm Policy, Economic Information Bulletin, Number 3, U.S. Department of Agriculture, June 2005, 1-2.

408 PT Bauer, Against the New Economic Order, Commentary Magazine, April 1977

INDEX

WND Books has a history of publishing provocative, current-events titles, including many *New York Times* bestsellers.

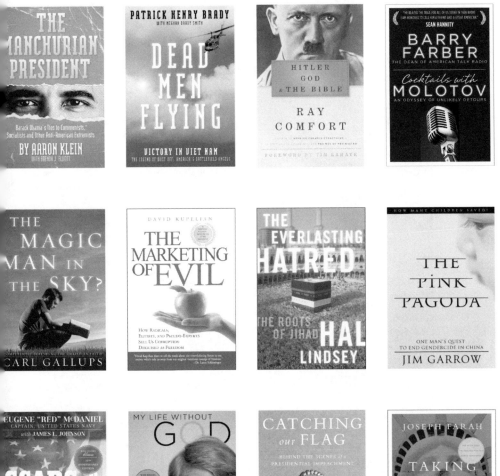

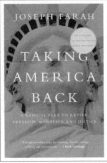

Sick of twisted "facts" mass-marketed to manipulate basic living decisions and common-sense energy consumption, Sussman indicts a cabal of elitist politicians, bureaucrats and activists who front the environmental movement to push intrusive, Marxist-derived policies in a quest to become filthy rich.

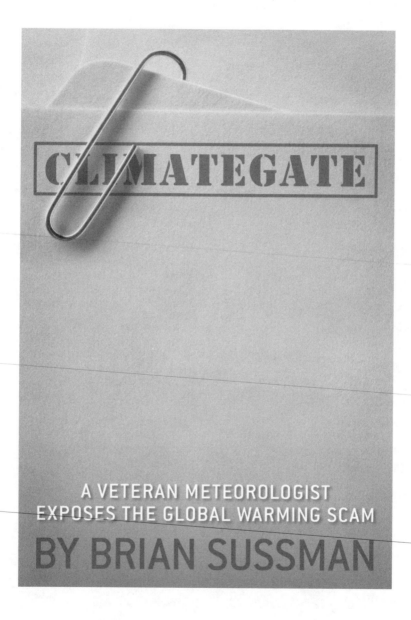

CLIMATEGATE

A VETERAN METEOROLOGIST
EXPOSES THE GLOBAL WARMING SCAM

BY BRIAN SUSSMAN

WND Books

WND Books • a WND Company • Washington, DC • www.wndbooks.com

Exorbitant energy prices, rolling blackouts, and acute food shortages: this is America's future as envisioned by the environmental movement's well-honed green agenda. *Eco-Tyranny* presents a rational strategy to responsibly harvest our nation's vast resources in order to fulfill the future needs of a rapidly growing population.

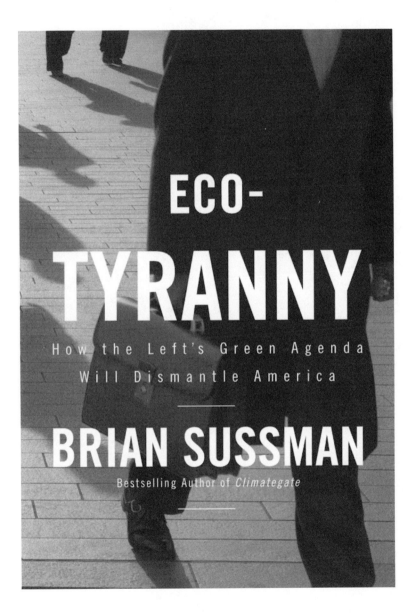

ECO-
TYRANNY

How the Left's Green Agenda
Will Dismantle America

BRIAN SUSSMAN

Bestselling Author of *Climategate*

WND Books

WND Books • a WND Company • Washington, DC • www.wndbooks.com